Danish Studies in Classical Archaeology
ACTA HYPERBOREA
9

Danish Studies in Classical Archaeology
ACTA HYPERBOREA
9

Pots for the Living
Pots for the Dead

Edited by
Annette Rathje, Marjatta Nielsen &
Bodil Bundgaard Rasmussen

MUSEUM TUSCULANUM PRESS
UNIVERSITY OF COPENHAGEN
2002

© 2002 Collegium Hyperboreum & Museum Tusculanum Press

COLLEGIUM HYPERBOREUM
c/o Institute of Archaeology and Ethnology
Vandkunsten 5, DK-1467 Copenhagen K.

Published by
MUSEUM TUSCULANUM PRESS
UNIVERSITY OF COPENHAGEN
Njalsgade 92
DK-2300 Copenhagen S
www.mtp.dk

Printed in Denmark
by AKA-PRINT A/S, Aarhus

Cover design: Thora Fisker
Decoration on base of Protocorinthian claybox. About 700 BC.

ISBN 87 72 89 712 0
ISSN 0904 2067

The publication of this volume has been made possible by generous grants from Carlsbergfondet and Konsul George Jorck og Hustru Emma Jorck's fond.

CONTENTS

TORBEN MELANDER & ANETTE RATHJE: Preface.......................... 7

HELLE SALSKOV ROBERTS: Pots for the Living, Pots for the Dead
Were Pots Purpose-made for the Funeral or Reused?
Can inscriptions throw light on the problem? 9

MARGIT VON MEHREN: The Trojan Cycle on
Tyrrhenian Amphorae... 33

TORBEN MELANDER: The Import of Attic Pottery to
Locri Epizephyrii. A Case of Reinterpretation...................... 59

BIRTE POULSEN: Genucilia –
Small Plate with a large Range 83

HELLE DAMGAARD ANDERSEN &
HELLE W. HORSNÆS: Terracotta House Models from Basilicata 101

ANNE MARIE CARSTENS: Archaic Karian Pottery –
Investigating Culture?... 127

LONE LEEGAARD: The Mediterranean and Central Europe in
the 6th and 5th centuries BC. The trade-route through
the Rhône Valley in the light of discoveries of local plain wares......... 145

KRISTINA WINTHER JACOBSEN: Cypriot Transport Amphorae
in the Archaic and Classical Period 169

JOHN LUND: The Ontogenesis of Cypriot Sigillata 185

ANNETTE GABRIELSEN SCHMIDT &
KAARE LUND RASMUSSEN: A Phiale in the
J.F. Willumsen Museum Collection – an Analysis of a Forgery 225

Forum

LONE WRIEDT SØRENSEN: The Archaic Settlement
at Vroulia on Rhodes and Ian Morris 243

Current Danish Classical Archaeological Fieldwork

ANNETTE RATHJE (ed.): Gazetteer of Danish Classical Archaeological
Fieldwork 1995-2001. With contributions by Søren Dietz,
Sanne Houby-Nielsen, Klavs Randsborg, Lone Wriedt Sørensen,
Maria Berg Briese, Anne Marie Carstens, Poul Pedersen,
Pia Guldager Bilde, Christina Trier 255

Book Reviews

ALEXANDRA NILSSON: Rev. of Fernando Rebecchi (ed.),
Spina e il Delta Padano 279

MARGIT VON MEHREN: Rev. of Cristina Chiaramonte Treré (ed.):
Tarquinia. Scavi sistematici nell'abitato. Campagna 1982-1988 287

HELLE W. HORSNÆS: Rev. of Renate Rolle und Karin Schmiedt (Hrsg.)
in Zusammenarbeit mit Roald F. Docter:
Archäologische Studien in Kontaktzonen der antiken Welt 292

Plates 1-23

PREFACE

The purpose of the publication *Pots for the Living, Pots for the Dead, Acta Hyperborea IX* is to present various contributions from a series of pottery-workshops. The first workshop took place at the Institute for Archaeology and Ethnology at Copenhagen University in 1995. The subject was *Greek Ceramics in the Mediterranean 800 -300 BC*. The participants were classical archaeologists from the universities of Aarhus and Copenhagen, and archaeologists from the collections of antiquities in the National Museum, Ny Carlsberg Glyptotek and the Thorvaldsen Museum. The lively debate at the workshop led to the decision that this should be a recurring event, that other topics in all areas of the Mediterranean world should be eligible topics for discussion, and that each workshop should make proposals for the following one. An organising committee was rapidly formed, consisting of Bodil Bundgaard Rasmussen (the National Museum), Torben Melander (the Thorvaldsen Museum) and Annette Rathje and Lone Wriedt Sørensen (both Copenhagen University). The topics for the subsequent workshops were as follows: *II Ceramics and Cult (1996), III Ceramic testimony on Cult and Ceremony in the Mediterranean 1st millennium BC (1997), IV Ceramics and Urbanisation (1998), V Ceramics. Complex Problem Structures and Problems (1999), VI Use and Meaning of Decorations on Ceramics (2000), VII Iconography and "In the Wake of Beazley"*.

Thanks to the generous support of the Carlsberg Foundation we were able to hold workshop III at the Danish Institute in Rome and V at the Danish Institute at Athens. Both places provided us with the opportunity to study ceramics on display in museums and in their storerooms. As a result of this collaboration, all the participants from the workshop participated in the Nordic debate that developed at the *Ceramics in Context colloquia* in Stockholm in 1997 and Helsinki in 1999.

All areas and all periods from the Iron Age to Late Antiquity in the Mediterranean world have, by now, been covered in the pottery workshops. The individual lecturers have, quite naturally, used Danish excavations as their point of departure, like those, for example, in Vroulia, Ficana, Pontecagnano, Castor and Pollux at the Forum Romanum, Aradippou on Cyprus, Panskoye on the Crimean peninsula, Chalkis in Aetolia, and Nemi. Other excavations, where Danes have participated, have also been presented, for example: Hymettos, Murlo and Knossos. Other locations have included Locri and Syracuse, Spina, Athens, Corinth and Sparta, whilst larger areas have been the Po valley and Central Europe, Campania and Caria. In addition to the main contexts - graves, sanctuaries and habitations, the contributions have mainly dealt with issues such as import, trade routes and local needs; gender, ceramics and social stratification, lifestyle, function analyses and iconography, ceramics and the reception of antiquities.

Torben Melander · Annette Rathje

POTS FOR THE LIVING, POTS FOR THE DEAD

Were pots purpose-made for the funeral or reused?
Can inscriptions throw light on the problem?

HELLE SALSKOV ROBERTS

A question often discussed by classical archaeologists is whether tomb gifts were purpose-made and specially bought for the funeral or were taken from things already available in the household.

A close look at some lines of a well-known inscription may elucidate the problem. The inscription is the cippus of the phratry of the Labyadae found at Delphi (Delphi Mus.inv.31: Rougemont 1977, 26-99, esp. 51-57). It contains a number of regulations concerning the funeral of the members of the phratry, all aiming at limiting extravagance in cost and untimely display of grief.

Lines 19-23 deal with *entophêia*, dialect form of *entáphia*, which literally means "things put into the tomb". It is stipulated that the value of these things, whether bought (*priámenon*) or "from the house" (*voíko*) should not exceed the sum of 35 drachmae, *i.e.* both possibilities are foreseen. The inscription then goes on to mention the mattress, cushion and cover to be used for the bier. Presumably these things are included in the *entáphia*. The fine for ignoring these regulations is 50 drachmae, *i.e.* more than one is allowed to spend on the tomb offerings in all.

The cippus is thought to be a copy from the first half of the 4th cent. BC of an archaic original from the end of the 6th or the beginning of the 5th cent. BC (Rougemont 1977, 42).

Bought for the funeral
As an example we may take the case of the white-ground lekythos ascribed to the Achilles painter (*CVA* Copenhagen, fasc.4, pls.170,6a-b; 170A; *ARV*² 997 no. 150) and the red-figured lekythos (*CVA* Copenhagen pl. 164,3; *ARV*² 1003 no. 18) thought by Beazley to be close

Fig. 1. White-ground lekythos by the Achilles painter from tomb at Keratia, Attica, in the Danish National Museum (Museum photo).

Fig. 2. Red-figured lekythos, close to the Achilles painter, from the same tomb as Fig.1, in the Danish National Museum (Museum photo).

Fig. 3. White-ground dummy lekythos from Chalkis, Euboia, in the Danish National Museum (Museum photo).

to the same painter. These two large lekythoi were found in the same tomb near Keratia in Attica. The fact that they were most likely acquired from the same workshop might support the idea that they were bought especially for the funeral, although, of course, there could be other reasons as, for example, the predilection of the deceased for the style of the Achilles painter and that the lekythoi were treasured by him during his lifetime. They are, however, extremely well preserved (**Figs. 1-2**) and they cannot have been put on to the pyre, the fate of many other, especially white-ground, lekythoi that are often broken in many fragments and show traces of a secondary burning.

Made for the funeral
Dummy lekythoi (e.g. *CVA* Copenhagen, fasc.4, pl. 172,1a-c) (**Fig. 3**) present a simpler case, being indubitably designed for exclusive use at the grave, and only capable of holding a minute measure of oil, quite in keeping with the spirit of the lawgivers seeking to limit the resources spent on the dead. Even so, they could be delicately decorated with scenes showing respectful visits to the grave. Probably these lekythoi were left standing on the steps of the tomb monument, informing the world of the living that the family had performed *tà nomizómena*, the proper rites. If this was the case the dummy was not strictly speaking an *entáphion*, not put into the grave, but used at rites performed after the burial, *i.e.* as illustrated by the frequent motif in vase-painting – and literature – "visit to the grave".

Rites performed at the grave
There are many examples in vase-painting of lekythoi and small vessels of other shapes standing at the steps of tombs, often adorned with wreaths. The Bosanquet painter has painted a particularly lavishly performed rite, showing numerous wreathed lekythoi and small jugs placed at the steps of a large monument (*ARV*² 1227 no.1) (**Fig. 4**).

The Penelope painter shows a single lekythos and a wreath put at the tomb of Agamemnon (*CVA* Copenhagen, fasc.8, pls. 351-352,1) (**Fig. 5**). Garland (1985, 115) suggests that the vases were smashed in the vicinity of the tomb, without, however, providing any evidence for such practice. Indeed, an inscription from Keos orders that the vessels should be brought back home (Sokolowski 1969 no. 97, lines 8-10): "Do not bring more than three choae of wine and one choe of oil to the tomb, and bring back the vessels".

Several vase paintings show lekythoi overturned on the tomb steps, which may be taken to indicate that they were left there after use (e.g. Kurtz 1975, pls. 23,3; 30,1) This would explain that they are often broken into many pieces, which has not necessarily happened deliberately.
The passage in Aeschylus, *Choephoroi*, vv. 94-99, sometimes thought to indicate that the vessels should be destroyed after use because of pol-

Fig. 4. Painting on white-ground lekythos by the Bosanquet painter in the National Museum in Athens (after Bosanquet 1899).

Fig. 5. Red-figured skyphos by the Penelope painter in the Danish National Museum (Museum photo).

lution (Mazon 1945, 83 n.2), on the contrary describes the abnormal situation felt by Electra, who is bringing the liquid offerings, *choae*, sent by Clytaemnestra. Electra feels that she cannot perform the usual rites because of the insincerity of the murderous giver of the choae.

Burial with modest entáphia. Tomb at Petróporos, Thessaly
As an example of a burial with modest grave-offerings of pottery, well within the limit of the 35 drachmae set by the Labyadae cippus, we may look at a tomb excavated at the village of Petróporos in Thessaly, which is thought to be the ancient Pelinna.[1]

In the tomb was a marble sarcophagus containing the skeleton of a woman and a bronze vessel holding a cremated child. On the lips of the woman was a gold coin with a head of a gorgon in mid-fourth century style. There was also a bronze coin issued by Antigonus Gonatas in the early years of his reign, which lasted from 277 to 239 BC.

The pottery included a lamp of the "inkwell" type, which was very popular from the third quarter of the 4th cent. BC into the later years of the century (**Fig. 6**). A close parallel to this lamp was found in a well-filling of the Agora of Athens, from the third quarter of the 4th century BC. Lamps were exported from Athens all over the Mediterranean area (Howland 1958, no. 238, Type 23 D, pls. 8;37).

Further pieces of pottery were:
Two black-glazed echinus bowls (**Fig. 7**). This is a very common shape which was produced in Athens from about 375 BC onwards and found at Corinth in the "Terracotta Factory", deposit 4, from the last years of the 4th century BC (Stillwell *et al.* 1984, 2335, pl. 126).

Two one-handled unglazed jars (**Fig. 8**); cf. Kurtz & Boardmann 1971, pl. 22 top row left, mid-4th century BC; also at Corinth, Stillwell *et al.* 1984, pl. 77, 2217, 2219).

A two-handled black-glazed cup on a ring foot (**Fig. 9**), cf. the cups nos. 1,2 and 8 from Hipponion tomb 19 mentioned below (Foti & Pugliese Carratelli 1974, figs. 3-6; 12-13).

Three open bowls covered inside and partly outside by black glaze of poor quality, on ring foot, two with incurving and one with outcurving rim.

The chronological range of the tomb gifts is defined by the coins from the middle of the 4th century BC to about 275 BC. The pieces of plain pottery are quite consistent with this period, although some of the shapes also occur earlier on. A period of 75 years, however, seems rather long to some scholars, who have commented on other aspects of this burial, and they suggest that the coin of Antigonus Gonatas and

Fig. 6. Lamp from tomb at Petróporos, Thessaly (Courtesy Dr. Tzafalias, Larisa).

Fig. 7. Two black-glazed echinus bowls from tomb at Petróporos, Thessaly (Courtesy Dr. Tzafalias, Larisa).

Fig. 8. Two unglazed jars from tomb at Petróporos, Thessaly (Courtesy Dr Tzafalias, Larisa).

Fig. 9. Black-glazed two-handled cup from tomb at Petróporos, Thessaly (Courtesy Dr Tzafalias, Larisa).

some of the pottery belong to the cremation burial of the child, the bronze vessel having been deposited inside the sarcophagus some time after the first burial of the woman (Tsantsanoglou – Parássoglou 1987, 4). However, the extremely exact indication of the position of the gold coin on the lips of the woman and of the two inscribed gold lamellae placed symmetrically on the chest at the time of excavation makes it less likely that the sarcophagus had been opened for a second burial. But the treasuring of a valuable gold coin for several years is by no means inconceivable and the types of plain pottery, which could also be used in the household, had a very long life. It was common practice in many periods in antiquity to have different forms of deposition for adults and small children at the same time. On balance, it is most likely that the two depositions were contemporaneous and made about 275 BC.

Fig. 10a-b. Ivy-shaped gold foils with inscriptions in Greek from tomb at Petróporos, Thessaly (after Tsantsanoglou & Parássoglou 1987).

Funeral offerings apart from pottery. Dionysiac symbolism
Outside the sarcophagus were found two terracotta statuettes, one very damaged, the other in the shape of a maenad, which, in the light of the contents of the inscribed gold lamellae, also found in the tomb, is hardly fortuitous.

When the burial was discovered in 1985 these gold lamellae naturally became the centre of attention, as they throw an interesting light on the religious beliefs of the deceased. They are ivy-shaped and made of paper thin gold foil, one measuring 40x31 mms, the other 35x30 mms. The text on both is the same, but the smaller is less complete. They are addressed to the person after death and instruct her to tell the goddess of the Underworld Persephone that Bacchios himself has released her and that therefore, presumably, she is qualified to join the other blessed in a new life (**Fig. 10a-b**).

Gold lamellae
Until now sixteen other gold lamellae containing written instructions to the deceased are known and they have been the subject of an animated discussion as to the precise religious groups they refer to, i.e. whether they should be termed Orphic or Bacchic.[2.]

Fig. 11. Polychrome terracotta statuette of maenad from tomb 934 at Lokri (after Ferri 1932-1933).

As Tsantsanoglou and Parássoglou (1987, 11) point out this is the first time Dionysos is explicitly mentioned ("the Bacchios who has released"), although the term *bacchoi* has been used on the other recently found gold lamella from South Italy, Hipponion tomb 19, where it refers to the acolytes of some cult, most likely that of Dionysos, but it does not refer specifically to Dionysos himself. The new Thessalian lamellae settle part of the discussion and justify the term Bacchic for these texts and tell us that the deceased lady was a participant in and, presumably, an initiate of the Dionysiac mysteries (Tsantsanoglou & Parássoglou 1987, 10-14). The statuette of the maenad found outside the sarcophagus should indubitably be seen as yet another piece of evidence for the woman's association with Dionysos and, possibly, it was her physical proof of the initiation.

Lokri tomb 934
A further indication of such a practice comes from tomb 934 at Lokri in South Italy, where the buried woman in her right hand was grasping a vase shaped as a remarkable polychrome maenad and, thereby, presumably marked out as an initiate of Bacchic mysteries (**Fig. 11**). In this

case the only other tomb-gift was a black-glazed cup with a lid decorated with a laurel wreath (Orsi 1913, 45-47).

Finds from Southern Russia
Such maenad-vases are not very common, but there are two parallels in the Hermitage (T 1852.52 and T 1852.5: Stephani 1869, nos. 2225 and 2230) (**Fig. 12**)[3]. They come from Sennaia, Phanagoria, in Southern Russia, where the influence of Dionysiac cult is noticeable, *e.g.* in Olbia, where some bone plaques from the 5th century BC with inscriptions referring to Dionysos have been found (Rusjaeva 1978, 87-104, figs. 1-6). The measurements are actually very similar to those of the gold lamellae. No.1: 5,1x4,1x0,2 cms; no. 2: 4,7x3,1x0,6 cms; no. 3: 4,8x3,5x0,5 cms. No. 1 has the graffito BIOC Θ.ANATOS BIOC, which might well be interpreted as referring to the rebirth to a new life also inherent in the first lines of the Thessalian lamellae: "now you have died, now you have been born…". On the same bone plaque is incised DIO ORΦIKOI and ALHΘEIA (truth) and in the middle is an A and, under this a sign, which is presumably a Z. This sign occurs once more on plaque 1 and also on nos. 2, 3 and 5 (**Fig. 13**). Rusjaeva suggests an interpretation of these signs as the numbers 1 and 7, possibly referring to the meaning given to these numbers in the Orphic symbolism and an application of them to the Dionysos myth, the Zeta-like sign and its variations on the bone plaques signifying the tearing to pieces of Dionysos by the Titans (Tinnefeld 1980, 69). This little-known story was, according to Pausanias, told in an epic by Onomacritus, an Athenian poet from the end of the 6th century BC (Paus. 8,37,5).

Further evidence of the cult of Dionysos from Olbia is a bronze mirror with the inscription DEMONACCA LENAIO EYAI KAI LENA OC DEMOKLO EIAY, naming Demonassa, the daughter of Lenaios and Lenaios, the son of Demokles. Lenaios has got his name from one of the epithets of Dionysos. EYAI and EIAY are the cult cries of the followers of Dionysos, known *e.g.* from the *Bacchae* of Euripides v. 157 (Rusjaeva 1978, 97 fig.7).

Herodotus also tells of how the cult of Dionysos spread among the Scythians, not, however, without meeting some resistance (Her. 4, 78-80).

Fig. 12. Polychrome terracotta statuettes of maenads from Sennaia, Panagoria (Courtesy Hermitage Museum).

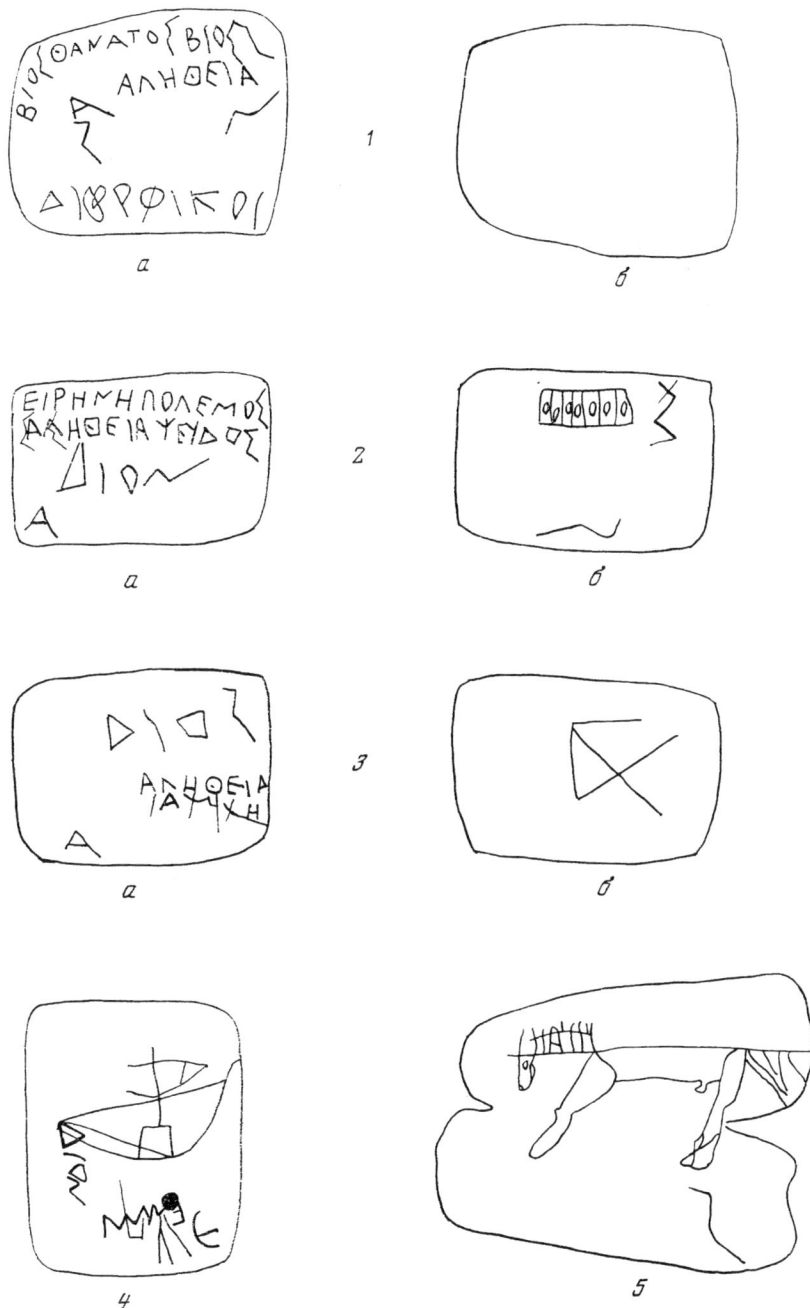

Fig. 13. Bone plaques from Olbia with graffiti (after Rusjaeva 1978).

Chamber tomb from Pharsalos in Thessaly
Another gold lamella with an interesting inscription comes from Pharsalos in Thessaly. It was found inside a bronze hydria, which was placed in a small stone chamber (1,20x1x0,85 m) (Verdelis 1950-1951, 81 fig. 1). A relief on the handle attachment figuring the rape of Oreithyia by Boreas, in Praxitelian style, quite reminiscent of the dancing maenad of the Lokri tomb (**Fig. 11**), points to a date of about 350-340 BC for the hydria and its contents, which, apart from the gold lamella, consisted in a plain bronze ring, diam. 1,3 cm, and a small two-handled skyphos with brownish glaze, which broke when touched (h. 4 cm; diam. of foot 5,5 cm; upper diam. 8,3 cm) (Verdelis 1950-1951, pls. 1-3).

The paper-thin gold foil (4,9x1,6 cm) is inscribed with nine lines in hexameter and deals with instructions to the deceased: "You will find in Hades's dwelling a spring to the right. Next to it stands a white cypress. Do not go near this spring. Further along you will find cool water flowing forth from the Lake of Mnemosyne. Above are guardians. These will ask you for what purpose you have come. To them you must tell the whole truth. Say: I am the child of the Earth and the star-studded sky. My name is Asterios. I am thirsty. Please give me to drink from the spring" (Verdelis 1950-1951, 99 fig. 11). This inscription does not mention Dionysos, nor does it refer to any initiation, but provides a magic formula which is supposed to gain access to the waters of Mnemosyne (**Fig. 14**).

The hydria from Pharsalos should be added to the list studied by Gisela Richter (1946, 361-367). She knew four or five hydriae with the motif of Oreithyia and Boreas and had four more groups with other, but related subjects, including Dionysos and Ariadne and Dionysos with a satyr, all of a rather homogenous style, which she dated to 350-320 BC, and she thought the hydriae to be the products of one workshop. She also discussed whether they were made for ordinary use or designed especially for funerary purposes. She concluded that the subjects were symbolical of love and marriage and that the primary use was to serve as water vessels, which occasionally could be used as cinerary urns. It is, of course, well-known that hydriae were often carried in wedding processions and presumably were part of the trousseau. Verdelis in his publication of the hydria from Pharsalos points to some specimens of alabaster which were obviously made for the tomb and

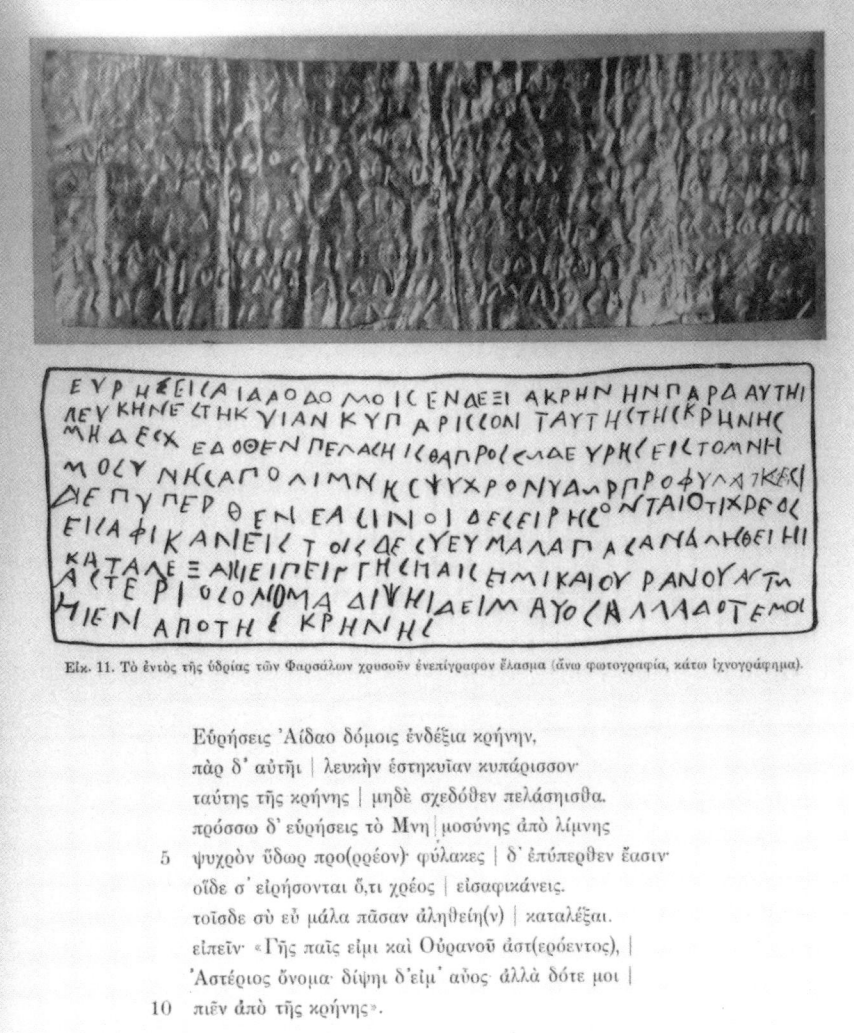

Fig. 14. Gold foil with Greek inscription found in hydria at Pharsalos (after Verdelis 1950-1951).

seems to favour the view that this was usually the case (Verdelis 1950-1951, 97-98).

The rich tombs of Macedonia, however, show examples of metal vessels used as cinerary urns, together with others that were obviously parts of table service given as tomb offerings (*Treasures of Ancient Macedonia* s.a., pls. 157; 184; 205; 338; 362).

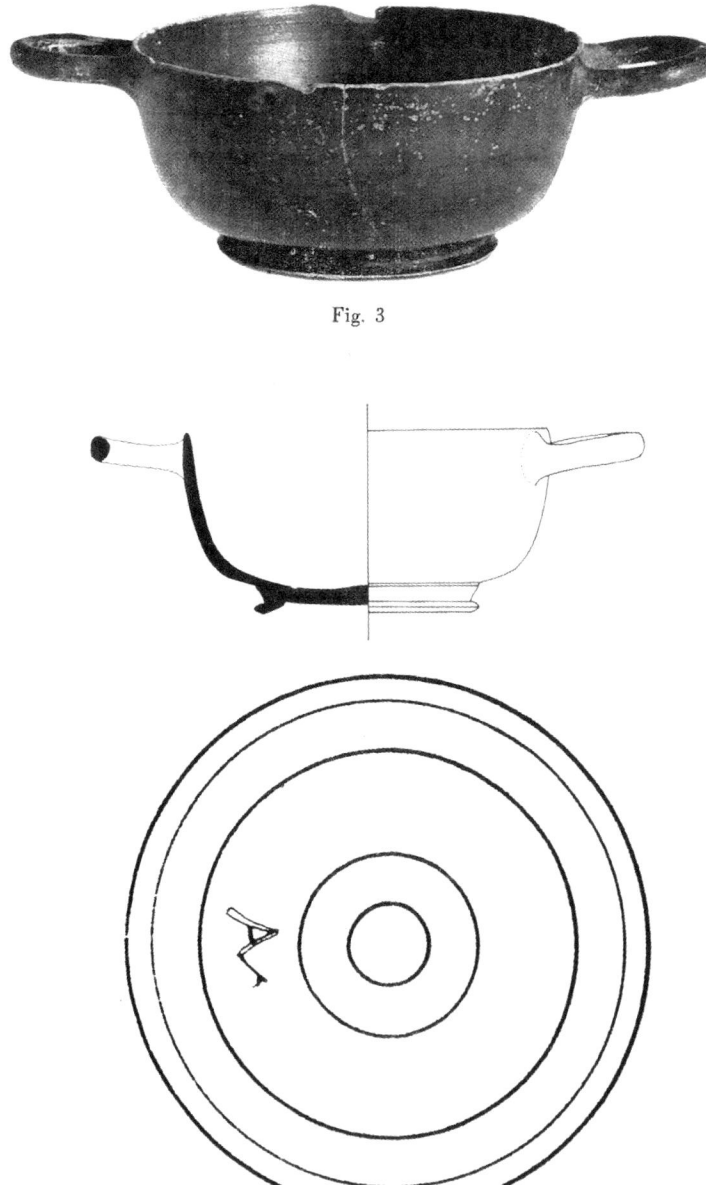

Fig. 3

Fig. 15. Black-glazed skyphos from tomb 19 at Hipponion with graffito under foot (after Foti & Pugliese Carratelli 1974. The drawing has been turned).

In this connection it seems worthwhile noticing that the Derveni tomb with the famous crater adorned with the elaborate Dionysiac reliefs also has a more direct relationship with the inscriptions on the gold lamellae, as scraps of a papyrus scroll inscribed with 18 columns from an Orphic (Bacchic?) poem were found in the remains of the offering pyre (Andronicos 1974, 271). This papyrus provides yet another piece of evidence as to how widespread the Dionysiac mysteries were in the latter part of the 4th century BC.

Hipponion tomb 19
With regard to South Italy the maenad from Lokri (**Fig. 11**) has already been mentioned and tomb 19 of Hipponion in Calabria has also been referred to. This tomb was *alla cappuccina* covered with tiles and on one of these tiles were found two low, black-glazed skyphoi (nos. 1-2) and a black-glazed one-handled lamp (no. 3) (Foti & Pugliese Carratelli 1974, figs. 3-6; 8) Under the foot of skyphos no.1 was a graffito (**Fig. 15**) which seems remarkably like the ones from Olbia (**Fig. 13, esp. 1a**).

But most unequivocally the inscribed gold foil (**Fig. 16**), which was found rolled up into a cylinder and placed on the chest of the well-preserved skeleton, proves the association of the deceased with a mystery cult similar to those documented by the now extensive list of gold lamellae, of which the ones from the Petróporos burial (**Fig. 10a-b**) and the gold foil from Pharsalos (**Fig. 14**) (Verdelis 1950-1951, pls. 1-3) have been discussed here.

The Hipponion inscription (5,9x3,2 cms) is a fuller version of the Pharsalos inscription on how to obtain the fresh water from the Lake of Mnemosyne (**Fig. 16a-b**). There is the same description of the white cypress standing by a spring to be avoided, the guardians of the Lake of Mnemosyne that will question the new arrival and the answer to give. But finally there is the promise of joining the other blessed *mystai* and *bacchoi* on the holy road. The Petróporos gold leaves (**Fig. 10a-b**) also refer to the other blessed (*olbioi*), but do not have the description of the Underworld and do not mention the Lake of Mnemosyne. They refer, however, also to an initiation of the deceased into the cult of Bacchios/Dionysos.

The two skyphoi and the lamp found outside the Hipponion tomb may show a final rite of libation performed after the closing of the tomb. The lamp may be taken to show that this also in Magna Graecia

Fig. 16a-c. Gold foil with Greek inscription from tomb 19, Hipponion (after Foti & Pugliese Carratelli 1974).

```
ΜΝΑΜΟΣΥΝΑΣΤΟΔΕΕΡΙΟΝΕΠΕΙΑΜΜΕΛΛΕΙΣΙΘΑΝΕΣΘΑΙ
ΕΙΣΑΙΔΑΟΔΟΜΟΣΕΥΕΡΕΑΣΕΣΤΕΠΙΔΣΙΑΚΡΕΝΑ            ΤΝ
ΠΑΡΔΑΥΤΑΝΕΣΤΑΚΥΑΛΕΥΚΑΚΥΠΑΡΙΣΟΣ                 ΑΟ
ΕΝΘΑΚΑΤΕΡΧΟΜΕΝΑΙΨΥΚΑΙΝΕΚΥΟΝΨΥΧΟΝΤΑΙ
5 ΤΑΥΤΑΣΤΑΣΚΡΑΝΑΣΜΕΔΕΣΧΕΔΟΝΕΝΓΥΟΕΝΕΛΘΕΙΣ
ΠΡΟΣΘΕΝΔΕΗΕΥΡΕΣΕΙΣΤΑΣΜΝΑΜΟΣΥΝΑΣΑΠΟΛΙΜΝΑΣ
ΨΥΧΡΟΝΥΔΟΡΠΡΟΡΕΟΝΦΥΛΑΚΕΣΔΕΕΠΙΥΠΕΡΘΕΝΕΑΣΙ
[.]ΟΙΔΕΣΕΕΙΡΕΣΟΝΤΑΙΕΝΦΡΑΣΙΠΕΥΚΑΛΙΜΑΙΣΙ
ΟΤΙΔΕΕΣΕΡΕΕΙΣΑΙΔΟΣΣΚΟΤΟΣΟΛΟΘΕΝΤΟΣ
10 ΕΙΠΟΝΥΟΣΒΑΡΕΑΣΚΑΙΟΡΑΝΟΑΣΤΕΡΟΕΝΤΟΣ
ΔΙΨΑΙΔΕΜΙΑΥΟΣΚΑΙΑΠΟΛΛΥΜΑΙΑΛΛΑΔΟΤΟ[
ΨΥΧΡΟΝΥΔΟΡΠ[..]ΡΕΟΝΤΕΣΣΜΝΕΜΟΣΥΝΕΣΑΠΟΛΙΜ[
ΚΑΙΔΕΤΟΙΕΛΕΟΣΙΝΙΥΠΟΧΘΟΝΙΟΙΒΑΣΙΛΕΙ
ΚΑΙΔΕΤΟΙΔΟΣΟΣΙΠΙΕΝΤΑΣΜΝΑΜΟΣΥΝΑΣΑΙΜΝΑΣ
15 ΚΑΙΔΕΚΑΙΣΥΧΝΟΝΗΟΔΟΝΕΡΧΕΑΗΑΝΤΕΚΑΙΑΛΛΟΙ
ΜΥΣΤΑΙΚΑΙΒΑΧΧΟΙΗΙΕΡΑΝΣΤΕΙΧΟΣΙΚΛΕΙΝΟΙ
```

took place before daybreak, as prescribed for Attica by the laws of Solon and by those of Plato for his ideal state (Rougemont 1977, 55).

Inside the tomb there was also a lamp placed in the left hand of the skeleton, this presumably meant to illuminate the path in Hades (Foti & Pugliese Carratelli 1974 no. 10, 101, figs 7; 15). On the pelvis was yet another black-glazed skyphos similar to the ones found outside the tomb (Foti & Pugliese Carratelli 1974, no. 8 figs. 12-13). On the fourth finger was a plain gold ring. One may recall that, correspondingly, the only tomb- gifts in the cinerary urn from Pharsalos were a small black-glazed skyphos and a bronze ring (Verdelis 1950-1951, 98). In the Hipponion tomb there were a few other pieces of plain pottery:

To the right of the head there was a one-handled unglazed jug, broken into many fragments (Foti & Pugliese Carratelli 1974, no. 4 fig.9). At the right elbow and at the left hand were unglazed miniature hydriskoi (Foti & Pugliese Carratelli 1974, nos. 6 and 11, figs. 11; 16). The jug, as well as the small hydriae, might well be thought to come in useful in carrying out the procedures in the Underworld with the two springs and a lake. Finally, there were some bronze fragments of a ring (or clasp ?) with an oval disc placed on the left shoulder and at the right elbow a small bronze hemisphere, perhaps part of a bell – to announce the new arrival? (Foti & Pugliese Carratelli 1974, nos. 5 and 7, fig. 10).

The Hipponion gold foil is dated before the other gold leaves to about 400 BC, based on the date of the pottery in the tomb, by scholars who have concentrated on the contents of the inscriptions. Perhaps it ought to be pointed out that the period during which such plain utilitarian pottery was produced and used could be rather long and that the Attic parallels referred to by Foti only provide a *terminus post quem*. None of the Hipponion specimens are said to be imports. For the type of fluted lekythoi nos. 12-13 Morel suggests a date in the second quarter of the 4th century BC (Morel 1981, pl. 167, 5414 b1) and the low skyphos represented by the three pieces nos. 1,2 and 8 continues into the last quarter of the 4th century (Morel 1981, pl.194, 6214 b1; Guzzo 1972, 545 fig. 13 no.2/14 from Praia a Mare, Cosenza). Bearing this in mind one cannot be too certain that the Hipponion tomb is much older than the Pharsalos hydria burial. Be that as it may, the fuller range of shapes represented, the careful observation of the placing of the objects and the existence of offerings outside the tomb, especially the skyphos with the graffito, make the Hipponion complex a find of the utmost importance.

Perhaps it is also worthwhile to look further north on the Italian Peninsula for evidence of Dionysos [4.]

Dionysos in Etruria
In Etruria there are several inscriptions referring to Pacha (Bacchus) including that of Laris Pulenas from Tarquinia (*TLE* ² 137; 190; 131) and from the same town comes the sculptured sarcophagus lid showing a woman with unequivocal Dinonysiac paraphernalia like the thyrsos, the kantharos and the young ram with reference to the tearing to pieces of Dionysos in the Dionysos/Zagreus version of his myth (Pryce 1931, 191-192 D 22, fig. 45).

Another Tarquinian sarcophagus, perhaps from the same workshop, has on the lid a male reclining figure holding out a phial for the kid to drink from, providing a remarkable illustration to *e.g.* the Petróporos text (Herbig 1952, pl. 27).

The French excavations at Bolsena/Volsinii have brought forward a large amount of plain black-glazed "Campanian" pottery extracted from a substantial layer of ashes, obvious evidence of a violent fire. The shapes represented include tall one-handled jugs, low skyphoi, lamps and plates. Pailler (1983, 11-39) suggests convincingly that these broken pots should be seen in the light of the well-known *senatus consultum "de Bacchanalibus"* of 186 BC. One original specimen of this decree issued by the Roman Senate was incised on a bronze plaque found in Tiriolo, Calabria, but, as Livy says, the sanctions against the cult of Bacchus were to be enforced all over Italy (Liv. 39,14,6-39,20). Livy states that the "contagious disease" which this cult seemed to be to the Roman Senate was believed to originate in Etruria, and indubitably the area of Volsinii would soon feel the efforts of the Roman magistrates to eradicate this threatening cult. The fragments of pots found in the ash layer are suggested by Pailler to be part of the cult equipment of Bacchic rites, although he regrets the absence of a definite, e.g. an epigraphical proof, that this burnt layer stems from the destruction of a Bacchic cult place.

But, perhaps, there is one, after all. The published material includes two graffiti that are obviously related, nos. 27 and 24, placed on the sides of two bowls (**Figs. 17a-b**). The graffitto on no. 27 seems to be letters of the Etruscan alphabet, while the one on no. 24 looks like a reversed and more formal version of the same signs. In both graffiti there is a clear A. The other signs are not immediately easy to interpret. If, however, one looks at the many-forked disrupted graffito on no. 27 combined with the A in the light of the signs on the Hipponion skyphos and especially the Olbia bone plaques, it is perhaps possible to see this as a version of the Zeta taken to be a symbol of the tearing to pieces, which seems to have played an essential part in Bacchic/Orphic cult (cf. Rusjaeva's interpretation of the signs on bone plaques **2b** and **3b** (**Fig. 13**) Tinnefeld 1980, 69).

Be that as it may, it is beyond doubt that the Orphic/Bacchic cult was well-established in Greece itself, from where it spread both eastwards to Southern Russia and to the Italian Peninsula in the West, from Calabria to Etruria, not without touching Rome, to the horror of the Senate, who acted to stop it in 186 BC. Perhaps it did not eradicate it totally, as a gold lamella with an "Orphic" inscription turned up in the

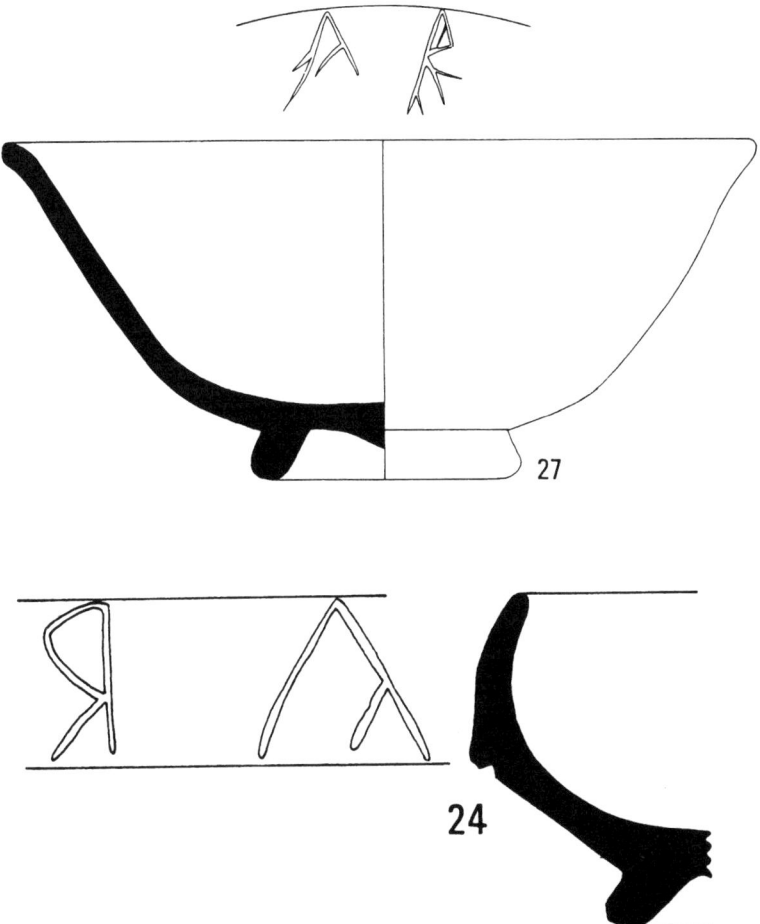

Fig. 17a-b. Two black-glazed bowls with graffiti from burnt layer at Bolsena (after Pailler 1983).

burial of Cecilia Secondina from the 2nd century AD (Zuntz 1971, 333 A5), showing a remarkable continuity of the tradition.

Reuse of Pots from the Household
Initiates of mystery cults like these were apparently not allowed spectacular tomb- gifts apart from the small gold foil with the formula essential to gain new life, but vessels like water- jugs, skyphoi and lamps in plainware seem to have been characteristic for these burials, presumably because they were thought to be important at various stages in the

Beyond and they may be indications of the beliefs of the buried people even when written evidence is missing. The funeral regulations of Delphi and Keos (Sokolowski 1969 nos. 77 and 97) show that it was often ordinary household vessels that were used for burials. These plain containers might be invested with symbolic meaning by the actual placing in the tomb or by the rites performed at the deposition.

NOTES

1: I am very grateful to the excavator Dr. Tzafalias, who has kindly provided photographs of the pottery found in the tomb and to the good offices of Dr. Aristea Papanicolaou Christensen.

2: See bibliography on the subject in Cole 1980, 223; Pugliese Carratelli 1993.

3: I am greatly indebted to Dr. Boriskovskaya and to Dr. Matveyev of the Hermitage Museum for permisson to illustrate these statuettes.

4: For fuller discussions, see *Orfismo* 1975; Cristofani & Martelli 1978; Pailler 1988; Nielsen 1990; Colonna 1991.

BIBLIOGRAPHY

Andronicos, M. & *al.* 1974
The Greek Museums. Athens.

ARV²
J.D. Beazley, *Attic Red-Figure Vase-Painters,* Oxford 1963.

Bosanquet, R.C. 1899
Some Early Funeral Lekythoi, *JHS 19, 169-184.*

Cole, S.G., 1980
New Evidence for the Mysteries of Dionysos, *GRBS* 21, 223-238.

Colonna G. 1991
Riflessioni sul Dionisismo in Etruria, *Dionysos, mito e mistero,* 117-155.

Cristofani, M. & M. Martelli 1978
Fufluns Paχies. Sugli aspetti del culto di Bacco in Etruria, *StEtr* 46, 119-133.

Dionysos, mito e mistero 1991
F. Berti (ed.) *Atti del Convegno Internazionale Comacchio 3-5 novembre 1989*. Ferrara 1991.

CVA. Copenhagen fasc. 4 and 8.

Ferri, S. 1932-1933
La menade di Locri, *BdA* 26, 262- 268.

Foti, G. & G. Pugliese Carratelli 1974
Un sepolcro di Hipponion e un nuovo testo orfico, *PP* 29, 91-126.

Garland, R. 1985
The Greek Way of Death. London.

Guzzo, P.G. 1972
Praia a Mare (Cosenza). Loc. Dorcara. Scavo di una necropoli del IV sec., *NS*, 535-548.

Herbig, R. 1952
Die jüngeretruskischen Steinsarcophage. Berlin.

Howland, R. H. 1958
Greek Lamps and their Survival (The Athenian Agora IV). Princeton.

Kurtz, D.C. 1975
Athenian White Lekythoi. Oxford.

Kurtz, D.C. & J. Boardman 1971
Greek Burial Customs. London.

Mazon, P. 1945
Eschyle, II. Les Choéphores. Paris.

Morel, J.-P. 1981
La céramique Campanienne: Les formes. Rome.

Nielsen, M. 1990
Sacerdotesse e associazioni cultuali femminili in Etruria: testimonianze epigrafiche e iconografiche, *ARID*, 19, 45-67.

Orfismo 1975
Orfismo in Magna Grecia, AttiMGrecia XIV. Napoli.

Orsi, P. 1913
Scavi, *NS* 1913 Suppl., 45-47.

Pailler, J.-M. 1983
Les pots cassés des Bacchanales, *MEFRA* 95, 7-54.

Pailler, J.-M. 1988
Bacchanalia. La répression de 186 av.J.-C. à Rome et en Italie: Vestiges, images, traditions, *MEFRA* 100, esp. 467 -.827.

Pryce, F.N. 1931
Catalogue of Sculpture in the Department of Greek and Roman Antiquities of the British Museum, I part ii. London.

Pugliese Carratelli, G. 1993
Le lamine d'oro "orfiche". Milano.

Richter, G. M. A. 1946
A Fourth-Century Bronze Hydria in New York, *AJA* 50, 361-367.

Rougemont, G. 1977
Corpus des inscriptions de Delphes. Paris.

Rusjaeva, A. S. 1978
Orfizm i kul't Dionisa v Ol'vii, *Vestnik Drevnei Istorii* 1978,1, 87-104.

Sokolowski, F. 1969
Lois sacrées des cités grecques. Paris.

Stephani, L. 1869
Die Vasensammlung der kaiserlichen Ermitage. St. Petersburg.

Stillwell, A. N. & al.
Corinth XV, part iii,. Princeton.

Tinnefeld, F. 1980
Referat über zwei russische Aufsätze, *ZPE* 38, 67-71.

TLE²
Testimonia Linguae Etruscae. Firenze 1968.

Treasures of Ancient Macedonia.
Exhibition Catalogue, s.l.e.a.

Tsantsanoglou, K. & G. M. Parássoglou 1987
Two Gold Lamellae from Thessaly, *Ellenika* 38, 3-16.

Verdelis, N. M. 1950-1951
Khalke tephrodokhos kalpis ek Pharsalon, *AÉphem* 1950-1951, 80-105.

Winter, F. 1903
Die Typen der figürlichen Terrakotten II. Berlin.

Zuntz, G. 1971
Persephone. Oxford.

Helle Salskov Roberts
Institute for Greek and Latin
University of Copenhagen
Njalsgade 90,
DK-2300 Copenhagen S,

THE TROJAN CYCLE ON TYRRHENIAN AMPHORAE

MARGIT VON MEHREN

In the following the focus is on the iconography of the Trojan Cycle on Tyrrhenian amphorae and on the way in which the Etruscans adopted, modified, and thus nationalized this Greek theme.

Among themes that serve as iconographical bases for the Attic so-called Tyrrhenian group of vases from the second quarter of the sixth century BC are two major ones belonging to legendary, mythological narrative cycles: that of Heracles and that of Troy.[1] They deal with events familiar to all civilized peoples around the Mediterranean, especially from the orally transmitted poems and tales, which soon became represented on finely wrought pottery. Aesthetic concerns must have been extremely important to the purchasers and users of these goods, and the narrative decorations must have been a prime consideration. The popularity of such thematic depictions is clearly attested to by the ubiquitous and numerous pottery finds which signal that abridged and adopted themes became pervasive in the sixth century when pottery in general, inspired by the Greeks, became the vogue in far-flung provincial centres. Still, Tyrrhenian vases are only one reflection of the profound acculturation from Greece in the Italic lands.

Tyrrhenian ware dealt with here furnishes a relevant and important example of the way in which Greek thematic treatments of glorious events were accepted and modified abroad. Purely numerically considered there is clear proof that the Heraclean feats soon became common folkloristic matter.[2] The legend of Troy, for example, achieved similar "international" status.

The largest number by far of Tyrrhenian vases of known provenance comes from excavations of Etruscan necropoleis during the nineteenth century, and none have been found in mainland Greece.[3] This serves to strengthen the commonly held view that the vases exclusively were made for export intended for Etruria. The shapes are Greek ones and typical of the time, but the iconography may very well have been chosen with an eye to prospective foreign buyers' taste, and it seems to be the case that certain subject matters became especially popular in Etruria. However, it should be remembered that it has recently been shown that this pottery was more widely distributed; isolated finds of Tyrrhenian ware have been identified in Asia Minor and the Near East.[4]

One interesting difference between export goods and domestic ones can be noted in the prevalence of lettered inscriptions and identifications on the former; it is as if foreigners were in greater need of elucidation and explanation: the foreign buyers were less familiar with these subjects than the Greeks. Furthermore, quite often on Tyrrhenian pottery we encounter nonsensical, Greek-"like" writing, which testifies to the idea that if something was "Greek-like" it might seem more attractive to foreign buyers who had little Greek. This or that object would be acceptable, provided that one might believe that the subject matter was indeed identifiably Greek – whether it really was so or not.

Frequent thematic variations and minor differences in executed details are extremely interesting because they show the great freedom enjoyed by individual painters of Tyrrhenian vases. The Tyrrhenian group is not a monolithic body, and there can be no doubt that individual painters shunned any suggestion of the stereotypical. This is confirmed by the liveliness of the depictions everywhere. These pots meant for export clearly have a highly dramatic narrative content; and so one might well wonder whether the painters were not deliberately catering to foreign predilections. At least this view gains credence when one examines the native Etruscan adaptations of the same scenarios.

There are eight Trojan subject matters represented on the Tyrrhenian vases: the Judgement of Paris, the Ambush of Achilles, the Pursuit of Troilos, the Fight for the Body of Troilos by Achilles and Hector, the Trojan Amazonomachie, the Duel of Achilles and Memnon, the Ransom of Hector, the Sacrifice of Polyxena.

The Judgement of Paris
In the *Iliad* (24.28-30) Homer briefly refers to this, one of the most favoured of events of ancient art; it is also mentioned in the *Kypria*.[5] The earliest known pictorial rendition is on the Protocorinthian Chigi Vase, where Paris stands waiting for the three goddesses who are led by Hermes.

The myth became popular on Attic black-figured vases from ca. 570 BC. The majority of the early renditions show another version which has its origin in Attica: the arrival of the goddesses at Mount Ida and Paris' stupefied turning-away as if in flight from such a grand epiphany (Raab 1972, 24-27).

On two Tyrrhenian amphorae, one in Havana the other in The Purrmann Collection we can note a similar, fairly sharply etched characterization.[6] They show the goddesses arriving one by one. As on other early black-figure they are dressed identically in peplos and mantle but without emblems; thus they become indistinguishable, their individuality lost. But later on, where they can be identified by clothing and attributes, the canonical order is Hera, Athena and Aphrodite. We may assume the same order on the early renditions. Only Hermes, the messenger of the Gods with his winged shoes, petasos, and kerykeion, is easily recognizable. The bearded Paris in his chiton and himation hurries away while looking back with his hand raised in a deprecating gesture as if declining the judicial task. On the amphora in Havana he holds a lyre, one of his herdsman's attributes. About this treatment of the subject Sparkes remarks:

"The three goddesses are no longer undifferentiated clones, and their epiphany has caused consternation in Paris who, like any sensible man in the circumstances, is disappearing fast, with a hasty glance back, sometimes with Hermes in hot pursuit."

Clearly such an initial reaction adopts an ironic, mock-heroic approach: Paris, the rustic country bumpkin, is overawed by such divine splendour and attempts his get-away with his cattle. On some vases later in the century we even see Hermes restraining and scolding the poor Paris.

But all things considered, the depictions of this scene on Tyrrhenian vases do not differ from those on remaining contemporary Attic black-figure. It is perhaps not too far-fetched to imagine that a mythological event like this could be treated irreverently. It had perhaps become such a commonplace that a nonchalant, ironic distance had become necessary.

From about the middle of the sixth century we know of two Etruscan depictions of the Judgement of Paris. One is seen on the "Boccanera Plaques" that once were mounted inside a tomb in the Banditaccia necropolis in Cerveteri.[7] Taken together, the three plaques constitute a continuous narrative frieze, which includes two small processions, one towards the left and one towards the right.

The leftmost procession may show the Judgement of Paris, as suggested by most scholars. If so, the bearded – as he is on early Attic black-figure – leftmost figure is Paris with a branch in his left hand, which may denote his oneness with nature and his homely slighted shepherd's trade. Towards him proceed the three goddesses led by Hermes wearing a petasos and with a sceptre with a figurine of a bullock as a finial that rests on his left shoulder. He has stopped before Paris. They are in animated discussion or exchanging salutations. Behind Hermes we see Athena holding a spear. The sequence continues on the next plaque with two female figures, each holding a pomegranate branch. The first is Hera who is followed by Aphrodite with her skirt hitched up and thus showing off her legs and her red shoes with pointed toes, the latest fashion from Ionia.

It is impossible to say whether the rightmost procession is related to the Judgement scene. It shows a tall long-haired woman, dressed beautifully and followed by three females with alabastrae and one with a pyxis, all containers for toilet requisites.

On a Pontic amphora in Munich the Judgement scene, which lent its painter the name "the Paris Painter", there is a lively rendition of the myth, which is enacted across both sides of the shoulder.[8] Paris is tending his father's flock on Mount Ida. A dog sits in front of three splendid bulls, the hindmost of which carries a raven-like bird on its back and watches the cattle, pointed ears and tongue hanging out; the young prince has turned away. He is leaning on his spear and raising his hand in greeting to the procession advancing towards him. Ahead of these figures walks an old white-haired, bearded man who in exactly measured kind returns Paris' greeting. Then follows Hermes with petasos. Both these heralds wear a kerykeion, and their presence concurs with the epic tradition. Hermes has turned to Hera and is apparently giving her last-minute directions while she listens eagerly. A mantle covers her head signalling her married state. Athena follows with a spear and hat or helmet, and the procession is concluded by Aphrodite with earrings, a ribbon round her neck and coyly raising her skirt; as on the Boccanera plaques, she shows her new shoes.

As in the case of Paris on the Tyrrhenian amphorae, the levitous treatment of the goddess shows that an august, universally known mythological event could be treated with humour.

The iconography of the Pontic amphora, which remains the only certain Etruscan depiction of the Judgement of Paris in the sixth century, seems to anticipate by some thirty or forty years the evolution of the motif on Attic black-figure; here it is not until late in the sixth century that Paris comes to be represented as a herdswain, and the canonical succession of the goddesses – Hera, Athena and Aphrodite cannot be established until then, as it can on the amphora of the Paris Painter from 550-540 BC (Raab 1972, 62; Kossatz-Deissmann 1994, 186).

The motif was employed later from the late Classical to early Hellenistic time – primarily but not numerously on mirrors (Kossatz-Deissmann 1994, 182, nos. 64-69, 188).

The Troilos Adventure
This can be divided into episodes: the ambush of Achilles at the fountain; the pursuit of Troilos; the killing of Troilos in the sanctuary of Apollo Tymbraios; and Achilles' and Hector's fight for the body of Troilos (Carpenter 1991, 17-21).

Primarily we know the legend, which especially was popular in the Archaic period, from depictions on pottery, for it received little written attention. The principal characters are Achilles and Troilos, and we know about Polyxena only from iconography.

The Ambush of Troilos by Achilles:
According to one version of the myth the Achaeans would lose the Battle of Troy if Troilos, son of Priamos, attained the age of twenty. Thus Achilles must kill Troilos before that birthday, and hence the Peleid waits in hiding by a fountain outside town. Here Polyxena comes to fill her hydria and so does Troilos to water his horses.

The subject is known from the *Kypria*, and the earliest known pictorial rendition is to be found on a relief-pithos from Tenos (ca. 675 BC). It became extremely popular on Attic black-figure but was widely employed on other pottery elsewhere as well in the Greek area during the Archaic period. On red-figure the motif is rare and gradually disappears entirely.

Tyrrhenian amphorae are among the earliest Attic black-figured pottery that depict the ambush; the theme is to be found on five amphorae (Mayer-Emmerling 1982, 77-79). The subject is rendered in its entire epic scope and with details that by and large agree with other, contemporary Attic black-figure.

Behind the fountain Achilles is concealing himself. His naked figure has hoplite equipment and his spear is raised for the attack. He crouches in the standard pose, shaded by a tree between himself and the fountain, e.g. on the Kiel amphora. On one amphora in Munich (**Fig. 1**) a warrior stands behind him. On another, in Philadelphia, he is on his feet, ready for the pursuit.[9] The fountain looks like a column from which is emitted a stream of water cascading from a pipe into the water basin in the shape of a low column on a squat base and with an Ionian capital. On the Munich amphora the water spills over on to the ground. Before the fountain stands Polyxena with a hydria soon to be filled.

An amphora in the Capitoline Museums in Rome exhibits some variations (Giglioli & Bianco 1962, 5, pl. 12,5-6; Kossatz-Deissmann 1981, 75 no.241). The tree is absent and the fountain is a square construction built of bricks with a spout formed as a lion's head. Here the water is also overflowing and Polyxena is bending forwards to fill her hydria from the streaming water. Atop the fountain sits the raven, Apollo's mantic bird, which later will inform his lord of the crime committed in the Tymbraic sanctuary. Next to Polyxena comes the mounted Troilos. He is naked, most often a boy or an adolescent. On the Rome and Munich amphorae he leads a spare horse in one hand. On the Munich amphora a prominent raven is in flight behind him. On the Kiel and Rome amphorae he is accompanied by one hoplite and on the amphora from Philadelphia by three.[10]

The Etruscans must have felt themselves attracted by this mythical subject: we see it in the second half of the sixth century on tomb painting, metal work, and vases;[11] but subsequently the motif disappears from Etruscan art.

The painting of the ambush in the Tomba dei Tori in Tarquinia has been considered the only known representation of a Greek myth in Etruscan tomb painting;[12] by and large it adheres to the traditional Greek iconographical scheme. As on the Tyrrhenian amphora in Philadelphia Achilles is on his feet ready to leap forth from behind the fountain, in front of which Troilos has halted to water his horse. But

Fig. 1. Tyrrhenian Amphora, München AS 1436. Photo Wehrheim.

here the fountain is a substantial structure of stone blocks in the form of an altar surrounded by trees – maybe laurels, Apollo's holy tree. Two lions rest between small trees or branches on top of it, one spouting water into a basin below. Between the fountain and the mounted

Troilos is a palm which, like the laurels, symbolizes that the event is taking place in Apollo's sanctuary.

On two bronze reliefs from the same matrix in the Villa Giulia Achilles is crouching behind the fountain with his long spear pointed at a young man holding a crock under running water emitted from the spigot-mouth of the lion. In the rear is Troilos with two horses.[13] The scene departs from Greek iconography by replacing Polyxena with the naked youth and by showing Troilos dismounted.

An Etruscan black-figure amphora, also in the Villa Giulia, once more shows Achilles in ambush behind the fountain.[14] The young Troilos is armed and approaches the fountain on foot. Behind him come his two horses followed by a warrior.

Two black-figure amphorae with several points of resemblance in Lucerne and the Vatican show a rendition of the myth that in several ways deviates from the Greek model.[15] The subject is unfolded on both sides of the vases. On one side Achilles is lurking behind the fountain, which is made of square blocks of stone and with a spout in the form of a lion's head. In its form the altar is reminiscent of that in the Tomba dei Tori. On the reverse Troilos is riding towards the fountain and leading another horse. This concurs with the standard Greek scheme. But the Etruscan vase painters imposed their own independent stamp on the episode. On the fountain stands a small, slight creature with its back to Achilles. It has long, erect, pointed animal ears and a trunk-like mouth. The feet resemble animal paws. In one hand it holds a knife or stick. On the Lucerne amphora the creature is placed in the middle of a grove which, as in the Tomba dei Tori, seems to grow above the fountain.

On these two amphorae a naked boy precedes Troilos with arms raised and a branch in his hand. But on the Lucerne amphora Troilos is not a youth; he is a mature, bearded man on horseback. Behind Troilos paces a tall warrior, also bearded, presumably Achilles, who seems to grasp at the horseman's head. The motif of these amphorae seems to be derived from an identical Etruscan model. On the Vatican vase Troilos naturally enough looks behind him, probably to keep an eye on his pursuer who may have been part of the depiction in the original composition.

Schauenburg interprets this as a representation of two episodes of the Troilos myth, distributed on either side of the amphorae, the ambush and the pursuit. In my view they are a combination, a conglomerate of two episodes that chronologically follow so hard upon one another that they are allowed to become one. This is especially clear on the Lucerne amphora, where the action must be read from left to right: Troilos rides towards the fountain behind which Achilles is hidden; at the same time the latter stands behind Troilos about to catch him. Such compressions are not unusual in Etruscan art (Schauenburg 1970, 71). Although we recognize the Greek framework at the heart of Etruscan iconography, each Etruscan artist left his own stamp on this event. Polyxena appears on no Etruscan fountain scene; the motif with the dismounted Troilos does not occur in Greece until the fourth century. The creature on the fountain is indeed inexplicable.[16]

The Pursuit
Achilles' pursuit of Troilos is a similarly popular subject, especially on black-figure. The subject is first represented on a Protocorinthian aryballos from ca. 700 BC. Later on it can be recognized on two Tyrrhenian amphorae, one in a Swiss private collection and one in the Louvre.[17]

The representation on the Louvre amphora shows the moment after the ambush. In complete hoplite armour with cuirass and leggings Achilles leaps forth from his hiding-place behind the fountain. From here he will pursue the offspring of Priamos in order to catch up with and kill him. Athena, his patroness during the Trojan War, stands behind him, identifiable by her attributes, the crested helmet, shield and spear. The fountain is shaped as a columnar post with the raven resting on top. A naked bearded man and Polyxena are in rapid flight looking backwards in fear and panic. The mounted Troilos is storming after them; the horses are frightened, their mouths open and eyes wild. The representation on this amphora is a variant, because Achilles still remains behind the fountain and is not in sharp pursuit as he normally is on contemporary vases.[18] This may indicate that on early Attic renditions there is still no sharp distinction between ambush and pursuit. Such a subtle distinction was yet to come.

On the Swiss amphora the subject is treated quiet traditionally as it is in most other renditions on Attic black-figure. Achilles is in front of the fountain pursuing Troilos who flees on horseback looking backwards at his pursuer. In front of Troilos runs Polyxena. She has lost her hydria, which now is on the ground below the horses. Athena stands behind the fountain on which the raven has lit.

In the second half of the sixth century the Etruscans depicted the pursuit and assault on their own vases and metal work. Subsequently the motif is continued on Etruscan red-figure as well as on mirrors; it also became popular on ash urns.

We know the subject from two Pontic amphorae, one in the Louvre and one in Reading. The Louvre amphora shows on one side of the shoulder Achilles who has caught Troilos by his hair and is tearing him from his horse.[19] A similar situation can be found on the Loeb tripod B in Munich. Here it looks as if Achilles were just about to kill the youth with his spear (Krauskopf 1974, 32, pl. 13; Camporeale 1981, 204, no. 63). This treatment is not known either from Tyrrhenian amphorae or from other black-figure, for here Troilos is invariably killed in the sanctuary, but it does adhere to Greek representations from the beginning of the fifth century and onwards.

The reverse of the Louvre amphora (**Fig. 2**) shows the pursuit, or the sacrifice, of Polyxena who attempts safety by climbing the altar-like fountain (there is no indication of a spout); but behind her stands Achilles with his spear pointed at her. He is followed by another warrior (Schauenburg 1970, 65, fig. 32; Hannestad 1976, 43-44). A depiction of the pursuit of only Troilos' sister is unknown on Greek vases where she and her brother are always in flight from Achilles. Accordingly the depiction on the Louvre amphora is an independent Etruscan version with no known Greek prototype.

Nor are there known models for the Reading amphora. On one side Achilles carries Troilos on his shoulder on the way to the altar. In other words, this is the moment just before he kills the youth in the sanctuary of Apollo Tymbraios.[20]

These Etruscan versions deal primarily with the moment when the son of Priamos has been caught and is thrown from his horse – a situation not known on Tyrrhenian amphorae.

Achilles' and Hector's Fight for Troilos' Body
This subject was far more rarely reproduced than were other episodes of the Troilos myth. Images of Achilles about to kill Troilos who has sought refuge on the altar in Apollo's sanctuary can be seen on shield-straps from Olympia in the first quarter of the sixth century. On a Cor-

Fig. 2. Pontic amphora, Louvre E 703. Photo Chuzeville. (Courtesy of the Museé du Louvre).

inthian crater of the same date the Trojans arrive while the killing is going on.

Depictions of the fight for the body are known from a few Attic black-figured vases, and the earliest and most dramatic ones occur on two Tyrrhenian amphorae, where the characters are identified by name-inscriptions (Mayer-Emmerling 1994, 79-83).

Troilos lies slain by Apollos' altar, which is in the form of an omphalos. On the Florence amphora he lies collapsed with his legs bent under him (Kossatz-Deissmann 1981, 87, fig. 360, 94-95). Achilles stands behind the altar, which by an inscription is identified as *Bomos*. He is helmeted and cuirassed with leggings, sword, spear, and a Boiotian shield. In his right hand he holds Troilos' decapitated head by the long hair, ready to fling it against the Trojans who, led by Hector, come to claim the body. They are naked but otherwise equipped as the Peleid. The only difference is that they carry circular shields and hold their spears ready for the attack.

Fig. 3. Tyrrhenian mphora, München AS 1426. Photo Koppermann. (Couresy of the Staatlischen Antikensammlungen und Glyptothek, München).

The Munich amphora (**Fig. 3**) contains a similar scene with but few departures.[21] Here Troilos' headless corpse lies flat on its back with knees lightly bent and hands contorted. Achilles is naked with his spear ready for fight and with Athena and Hermes behind him. In the air between Achilles and Hector floats the decapitated head. Either the Peleid has thrown it at Hector or he has speared it and is provocatively displaying it to his enemy.

The myth of Troilos, with the principal emphasis on the ambush and the pursuit, was heavily favoured in art, especially in the sixth century. This may be accounted for by the fact that Troilos' death made the Greeks' victory possible.

In Etruria Troilos' death and the fight for his body are known exclusively from a number of Hellenistic ash urns.[22]

Nevertheless, I believe that the subject occurs earlier on a Pontic amphora in Heidelberg (**Fig. 4**).[23] Here we see the fight of two warriors. One is bursting forward, spear raised, while the other is retreating and

Fig 4 a. Pontic Amphora, Heidelberg inv. 59/5. (Courtesy of the Antikenmuseum des Archäologischen Instituts der Universität Heidelberg).

Fiv. 4 b. Detail of fig. 4a. (Courtesy of the Antikenmuseum des Archäologischen Instituts der Universität Heidelberg).

covering himself with his shield; yet he is already defeated; his knees are buckling under him and he is mortally wounded with blood streaming from his throat. Fronting him is another warrior. Above and between the combatants is a three-winged head. The depiction has a resemblance to those on Tyrrhenian amphorae that show Achilles throwing Troilos' decapitated head at Hector. But in this version Hector has already been defeated, and not until now has the head been thrown; now it is afloat between the duellists and thus there is kinship to the Tyrrhenian amphora in Munich. The Etruscan vase-painter chooses to show the head in flight by applying wings to it.

The Duel of Achilles and Memnon

This myth was told in the now lost *Aithiopis*. Memnon, King of the Ethiopians and son of Eos and Tithonos, was an ally of Priamos in the Trojan War. As we know, Antilochos, friend of Achilles, was killed by Memnon; this in turn led to his death by the hand of Achilles and in conformance with Zeus' *kerostasie*. A similar treatment of events occurs on the Hydria Ricci where we se the duel as well as the *kerostasie*, the weighing of the duellist's souls (d'Agostino & Cerchiai 1999, 134-138, figs. 77-78). But the duel per se of heroes became a favoured theme in Archaic art and especially in Attic black-figured vase-painting. It was not until the fifth century that the subject ceased to be of interest.

A firmly-established, stereotype scheme emerged showing a duel between two warriors. The fallen Antilochos, who was the cause of the fight, may lie between them. In many versions the defeat of Memnon has already been anticipated. Most often this scene is framed by the two divine mothers, Eos and Thetis, each imploring Zeus for her son's victory. There may be secondary characters, onlookers and horsemen.

One cannot immediately identify a duel with (or without) someone fallen and two women present as the duel between Achilles and Memnon; only written identifications of the characters on the vases can make it certain that such scenes represent that particular myth.

The earliest preserved rendering furnished with identifying inscriptions is that on a Corinthian crater from Cerveteri, depicting an indefinable heroic duel flanked by two horsemen, from 580 BC, a time when this subject was often used.

The motif can be recognized on Attic black-figure from ca. 570 BC, primarily on Tyrrhenian amphorae.[24] There are at least twenty amphorae depicting the scene, one of which is in Cerveteri, with written identification of Chaicos, Eos, Memnon, Achilles, and Diomedes.[25] On another amphora in New York the hard-pressed Memnon has been brought down on his right knee and Achilles, identified by an inscription, faces him on the right. In the background the mothers gesture intensely and apprehensively (Moore & von Bothmer 1976, 2-4, pl. 3). These two are the only examples with "proper" identification.

The renderings on the rest of the Tyrrhenian amphorae also follow the prototype and therefore probably show the same subject. Two of these vases – one in Leiden and another in Frankfurt – on which the combatants are fighting over the body of a fallen warrior – clearly allude to the myth. Not infrequently the figures are accompanied by nonsense inscriptions.[26]

There is a peculiar, unique presentation of the duel on an amphora in the Louvre (**Fig. 5**). Here we see a duellist who has been pierced by his opponent's sword, and forced to his knees, looks behind him. The victor who looms large holds a Boiotian shield covered by an animal hide, perhaps appropriately that of a crocodile. For he is Memnon, King of the Ethiopians and Achilles is the loser. The presence of two women holding garlands suggest that the subject is indeed the Achilles-Memnon duel, even though the outcome is not what one might expect (Pottier 1922, 5, pl. 5,3).

In the Etruscan Collection in Basle there is a Pontic amphora (Simon & Hampe 1964, 22-25, pls. 8, 11; Hannestad 1976, 13-14), which I believe depicts the duel between Achilles and Memnon and adheres to the Greek prototype. This fight is flanked by the goddesses, the heroes' mothers. Although there are two women behind the leftmost warrior, these are clones and such a reduplication of divinities is not unusual (Schauenburg 1970, 39).[27]

Furthermore on the bronze chariot from Monteleone, now in New York, there is a duel over the body of a slain warrior.[28] No females are present. The duellists have been interpreted as Achilles and Memnon, but the subject might just as well be an unidentifiable fight.

Fig. 5. Tyrrhenian amphora, Louvre E 854 Photo Lebaube. (Courtesy of the Musée du Louvre).

THE TROJAN CYCLE ON TYRRHENIAN AMPHORAE

The Trojan Amazonomachia

On a number of Tyrrhenian amphorae are depicted battles between Greek warriors and Amazons.[29] They might well represent the Achaeans fighting the Amazons who were allied with and assisted the Trojans, as described in the *Little Iliad*. Myth tells us that Achilles killed Penthesilea, Queen of the Amazons in a duel.

These amphorae often display typical duel scenery with the Greek hero having forced an Amazon to her knees. Thus this rendition is analogous with presentations of Heracles' fight with Andromache. But in the absence of identifying inscriptions, such scenes might just as well depict Greeks battling Barbarians; certain identification is difficult; it might be remembered that on a slightly later amphora by Exekias the same battle plan was furnished with the written names of Achilles and Penthesilea. This version may have provided the basis for interpreting any and all unidentified battle scenes between Greeks and Amazons but with Heracles absent as dealing with the Trojan War.

The Ransom of Hector

In the *Iliad* (24, 469-491) Homer carefully describes the visit of King Priamos to Achilles to whom he bears a multitude of costly gifts and pleads for the release of his son's body for proper burial. The earliest depictions of the myth are known from Peloponnesian metal work from 570-560 BC. A few contemporary renditions occur on Attic black-figure, which from the outset adhere to a traditional scheme later continued on subsequent black- and red-figure. The motif seems not to have been widely employed among the Greeks.

One of the earliest depictions of the story is found on a fragmentary Tyrrhenian amphora from Cerveteri in the Louvre.[30] The bearded Achilles is reclining on a couch in the new oriental manner adopted in both Greece and Etruria about 600 BC. He has a drinking bowl in his right hand and his weapons hang on the wall. In front of the kline is a table and before that Hector's body. In supplication Priamos is approaching the foot of the kline with hasty steps, and his face seems to reflect grief and despair. The Trojan king is depicted as an old man with white hair and beard. Behind Priamos a man and a woman converse. The woman is lifting her mantle from her face. The figures might be Briseis and one of Achilles' servants, Automedon or Alkimos. The subject depicted on the reverse is possibly an extension of the motif, for it shows some elderly men turning towards a large tripod. A flute-player stands with his back to it; this might well be Priamos' escort bringing a part of the ransom.

We know from the detailed description in the *Iliad* that Priamos was not to see his son's body until the next morning when it was placed, beautifully garbed, on the cart that had brought the ransom. Thus the pictorial renditions differ from the epic narration. Two closely related sequences in Homer are for economy's sake compressed into one elaborate illustration of the myth.

As far as we know, Hector's ransom is not a subject of Etruscan art.

The Sacrifice of Polyxena
According to the *Iliupersis*, Priamos' daughter Polyxena was sacrificed at Achilles' tomb after the fall of Troy.

On a Tyrrhenian amphora in London there is a unique presentation of the event, in which the characters are clearly identified by inscriptions.[31] The dramatic action is cynically represented with all its brutal and bloody terror. We are in the midst of the sacrificial act. The young Trojan princess is so firmly wrapped in her elaborately woven peplos that she mostly resembles a bundle, with her arms strapped as well in order to prevent her moving (Conelly 1996, 61-63). Like a sacrificial animal she is held horizontally by three hoplites with her face downwards over Achilles' tomb. Amphilochos has firm hold of her bosom, Antipathes holds her by her knees and Aias Iliades controls the feet. Behind the tomb the sacrificial fire is lit on the altar. Neoptolemos steps forcefully before her and holds up her head by her hair with his left hand and with the other he slits her throat with his sword. A torrent of blood courses from the throat upon the tomb. Behind Neoptolemos, Diomedes and Nestor Pylios are interested onlookers. Behind Polyxena the old Phoenix has gotten to his feet and averts his eyes from the cruel sight.

This is the sole known and certain depiction of the sacrifice itself.[32] Later vases represent Polyxena being led to the tomb.

In Etruscan art we know a single representation of the killing of a woman, probably Polyxena, beside a tomb. The scene is found on the sarcophagus from Torre San Severo from the late fourth century B.C. The spirit of the departed hero is watching the ceremony as he leans on his tombstone (Touchefeu-Meynier 1994, fig. 32).

Conclusions and remaining questions

To isolate the Trojan events from the wealth of other iconographical matter on the Tyrrhenian amphorae is one way of applying a magnifying glass to a special area of vasepainting. From it we can realize what were special focuses of the painters and their foreign customers. But some inescapable questions are left: why for example did the painters not in the Trojan accounts pay attention to the Armour of Achilles or to the Sack of Troy? Why is the focus so clearly on these eight events only?

It might be possible to come to more far-reaching although still tentative answers about these Etruscan modifications of narrative contents and emphases if the magnifying glass were shifted for example to the Theban Cycle or to the tales of the Gods.

Depictions of events from the Trojan Cycle were known and beloved all over the Greek world both in the Archaic and the Classical periods. But Tyrrhenian amphorae only show a select choice of these events with their emphasis on blood, murder, human sacrifice and the *causa bellis*. There are neither representations of love nor of heroism. In his adventure with Troilos Achilles has been made a murderer of a youth, and – the Judgement of Paris apart – the Tyrrhenian iconography of the Trojan cycle conveys almost nothing but feelings of horror and fear.

Most of these Trojan subjects were adopted by the Etruscans during the sixth century. But only in a few of the renderings are we able to point to the original prototype. The Etruscans transformed the motifs in their own way and modified, omitted, or added details.

The Etruscan artists probably knew the tales from both the oral tradition and from Greek renderings. The many hybrid Etruscan depictions of the Trojan events – at least as bloody and horrible as the Tyrrhenian ones – point to this solution.

NOTES

1. Mayer-Emmerling 1982. In an article (von Mehren 2001) I deal with the iconography of the Heracles myth.

2. Approximately 260 Tyrrhenian vases are known today, which depict secular or mythical Greek events. The motifs are genre depictions and Greek myth. The mythical heroic deeds are 100 in number: Theseus killing the Minotauer 3, Perseus pursuing the Gorgons 3, the hunt for the Caledonian boar 7, the Theban cycle 6, the Trojan cycle 35, Heracles 46. The total number of the tales of the Gods is 26.

3. Thiersch 1899; Kluiver 1992, 1993, 1995 and 1996

4. Tuna-Nörling 1997. All known Tyrrhenian pottery shapes (amphora, hydria, krater, dinos, oinochoe and plate) belong to banquet equipment; but not one single drinking-cup has been found.

5. As we know, the *Iliad* deals only with one of the ten years of the Trojan War. Other events were related in other, post-Homeric poems as the *Kypria*, the *Aithiopis*, the *Iliupersis* and the *Little Iliad*. These were transmitted only fragmentarily and later summarized by Proclos (probably AD 412-485).

6. Both: Raab 1972; Mayer-Emmerling 1982, 74-77. Havana: Kossatz-Deissmann 1994, 178, fig. 9. Purrmann Collection: Schauenburg 1973, 25, figs. 29-32.

7. E.g. Murray 1889, 243-252; Roncalli 1965, 69-77, pls. 12-15; Haynes 1976, 227-231, pl. 69; Kossatz-Deissmann 1994, 180, no. 41. The plaques were reemployed in the tomb; probably they came from a dwelling house.

8. Ducati 1932, 7-9, figs. 1-2; Hannestad 1974, 17, figs. 1-2; Kossatz-Deissmann 1994, 180, no. 42.

9. Kiel: Schauenburg 1970, 54 fig. 21; Kossatz-Deissmann 1971, 75, no. 223. Munich: Kunze-Götte 1970, 13-14, pls. 313, 1-2, 314, 1-2; Kossatz-Deissmann 1981, 75 fig. 230. Philadelphia: Schauenburg 1970, 47 fig. 11; Kossatz-Deissmann 1981, 76 fig. 238.

10. The fifth amphora is in the Vatican Museum: Kossatz-Deissmann 1981, 76 no. 240, not illustrated.

11. This subject may occur already in the late seventh century on a bucchero amphora with relief decoration from Cerveteri (Prayon 1977, 181-197, pl. 95) and in the beginning of the sixth century on an Etruscan amphora (Kossatz-Deissmann 1997, 92 no. 3, not illustrated).

12. Banti 1955-1956, 143-154; Giuliano 1969, 3-26, pl.1; Simon 1973, 27-42, fig. 1; Oleson 1975, 189-200; Cerchiai 1980, 25-39; Camporeale 1981, 210; d'Agostino 1985,

1-8, d'Agostino & Cerchiai 1999, 91-114. Apparently this view is untenable if the Boccanera plaques in fact depict the Judgement of Paris (note 6).

13. Bronze plaquette, Tomba del Guerriero, Vulci: Camporeale 1969, 65-70, pl. 27; Camporeale 1981, 202, fig. 15.

14. Camporeale 1969, 70-74, pl. 28; Schauenburg 1970, 62-63, figs. 29-30, Camporeale 1981, 202, no. 13.

15. Tolfa Group, 540-530 B.C.: Schauenburg 1970, 68-80, figs. 36-37, fig. 38; Camporeale 1981, 202 nos. 11-12; d'Agostino & Cerchiai 1999, 115-119.

16. For interpretations: Banti 1955-1956, 38-40; Schauenburg 1970, 74-81; d'Agostino & Cerchiai 1999, 115-119.

17. Swiss private collection: Kluiver 1996, 2, no. 12, fig. 5. Louvre: Del Chiaro 1964, 108; Krauskopf 1977, 33, fig. 16-17; Mayer-Emmerling 1982, 79.

18. A similar scene occurs only once on a contemporary Siana kylix by the C-painter, now in the Louvre: Kossatz-Deissmann 1994, 83, fig. 310.

19. Schauenburg 1970, 64, fig. 31; Krauskopf 1974, 32, pl. 14,2; Hannestad 1976, 43; Camporeale 1981, 202 no. 17, 210-211.

20. Schauenburg 1970, 66-68, fig. 34; Hannestad 1976, 27-28; Camporeale 1981, 202, fig. 18.

21. Kunze-Götte 1970, 11-13, pls. 312, 313,2-3; Kossatz-Deissmann 1981, 87, fig. 364.

22. Nielsen 1993, 322-323, motif nos. 13, 14, 21, 328-329, fig. 1.

23. Hampe & Simon 1964, pl. 1; interpreted as the duel between Herales and Kyknos.

24. Kunze-Götte 1970, 22-24, pl. 321; Rastrelli 1981, 10, pls. 7-8; Mayer-Emmerling 1982, 87-89.

25. Beazley 1971, 40; no illustration is available.

26. Leiden: Jongkees-Vos 1972, 3-4, pl. 3; Tiverios 1976, pl. 19a. Frankfurt: Deppert 1964, 27-28, pl. 24. On an amphora in Karlsruhe showing the canonical scheme a naked corpse is seen in front of the combatants. Two warriors have raised it from the ground by its hands and feet, ready to carry it away – or they pull in opposite directions in order to take possessions of the body. Thiersch suggests this scene as the fight for the body of Patroclos. Thiersch 1988, 65 no. 58, pl. 6; Hafner 1951, 15, pls. 5-6. I think it may show an unique representation of the duel between Achilles and Memnon.

27. Hampe & Simon interpret the scene as Tydeus' fight with Polyneices.

28. Montoleone chariot: Simon & Hampe 1964, 57-60, pl. 24; *Carri da guerra* 1999, 179-190, frg. 6 left.

29. von Bothmer 1957, 10-11, 24-27, pls. 20-25; Mayer-Emmerling 1982, 84-86.

30. Lullies 1964, 82, pl. 28,3-4; Mayer-Emmerling 1982, 90-92; Kossatz-Deissmann 1994, 148-161, no. 647. The rendition had been extensively restored and painted over, which caused misinterpretations (e.g. Holwerda 1890, 242, fig. 17).

31. Walters 1898, 285; Mayer-Emmerling 1982, 93-95; Touchefeu-Meynier 1994, 431-435, fig. 26.

32. On a fragmented Proto-Attic krater in Boston (ca. 650-630 BC) is a depiction of a woman being carried in the same manner, as is Polyxena on the London amphora. This has been interpreted either as the sacrifice of Polyxena or of Iphigenia; Vermeule & Chapman 1971, 285-293.

BIBLIOGRAPHY

Banti, L. 1955-1956
Problemi della pittura arcaica etrusca: la Tomba dei Tori di Tarquinia, *StEtr* 24, 145-154.

Beazley, J.D. 1971
Paralipomena. Oxford.

Bothmer, D. von 1957
Amazons in Greek Art. Oxford.

Camporeale, G. 1969
Banalizzazioni etrusche di miti greci III, *StEtr* 37, 65-70.

Camporeale, G. 1981
Achle, C.I. Achilleus e Troilos: agguato, *LIMC* I, 202-211.

Carpenter, T.H. 1991
Art and Myth in Ancient Greece. London.

Carri da guerra (I) 1999
I carri da guerra e principi Etruschi. Exh. cat. ed. A. Emiliozzi. Viterbo 1997-1998, 2. ed. Roma 1999.

Cerchiai, L. 1980
La machaira di Achille: alcune osservazioni a proposito della Tomba dei Tori, *AION* 2, 25-39.

Conelly, J.B. 1996
Parthenon and *Parthenoi*: A Mythological Interpretation of the Parthenon Frieze, *AJA* 100, 53-80.

d'Agostino, B. 1985
Achille e Troilo. Immagine, testi e assonanze, *AION* 7, 1-8.

d'Agostino, B. & L. Cerchiai 1999
Il mare, la morte, l'amore. Gli Etruschi, i Greci e l'immagine. Roma.

Del Chiaro, M.A. 1964
Classical Vases in Santa Barbara, *AJA* 68, 107-112.

Deppert, K. 1964
CVA Deutschland, Frankfurt am Main I. München.

Ducati, P. 1932
Pontische Vasen. Berlin.

Giglioli, G.Q. & V. Bianco 1962
CVA Italia, Musei Capitolini I. Roma.

Giuliano, A. 1969
Osservazioni sulle pitture della Tomba dei Tori, *StEtr* 37, 3-26.

Hafner, G. 1951
CVA Deutschland, Karlsruhe, Badisches Landesmuseum I. München.

Hannestad, L. 1974
The Paris Painter. Copenhagen.

Hannestad, L. 1976
The Followers of the Paris Painter. Copenhagen.

Hampe, R. & E. Simon 1964
Griechische Sagen in der frühen etruskischen Kunst. Mainz.

Haynes, S. 1976
Eine etruskisches Parisurteil, *RM* 83, 227-231.

Holwerda, A.E.J. 1890
Korinthisch-Attische Vasen, *JdI* 5, 237-268.

Jongkees-Vos, M.F. 1972
CVA The Netherlands, Rijksmuseum van Oudheden I. Leiden.

Kluiver, J. 1992
The "Tyrrhenian" Group: its origin and the neck-amphorae in the Netherlands and Belgium. With a contribution on petrography by H. Kars, *BABesch* 67, 73-109.

Kluiver, J. 1993
The potter-painters of the Tyrrhenian neck-amphorae: a close look at the shape, *BABesch* 68, 174-194.

Kluiver, J. 1995
Early Tyrrhenian: Prometheus Painter, Timiades Painter, Goltyr Painter, *BABesch* 70, 55-103.

Kluiver, J. 1996
The five later "Tyrrhenian" painters, *BaBesch* 71, 1-58.

Kossatz-Deissmann, A. 1981
Achilleus. VII Das Troilosabenteuer, *LIMC* I, 72-95.

Kossatz-Deissmann, A. 1994
Paridis Iudicium, *LIMC* VII, 176-188.

Kossatz-Deissmann, A. 1997
Troilos. Achilleus lauert Troilos auf, *LIMC* VIII, 92.

Krauskopf, I. 1974
Der thebanische Sagenkreis und andere griechische Sagen in der etruskischen Kunst. Mainz.

Krauskopf, I. 1977
Eine Attish Schwarzfigurige Hydria in Heidelberg, *AA*, 28-37.

Kunze-Götte, E. 1970
CVA Deutschland, Museum Antiker Kleinkunst VII. München.

Lullies, R. 1964
Eine Amphora aus dem Kreis des Exekias, *AK*, 82-90.

Mayer-Emmerling, S. 1982
Erzählende Darstellungen auf "Tyrrhenischen" Vasen. Frankfurt.

Mehren, M. v. 2001
Two Groups of Attic Amphorae as Export Ware for Etruria: the so-called Tyrrhenian Group and Nikosthenic Amphorae, *Ceramics in context. Proceedings of the Internordic Colloquium on Ancient Pottery held at Stockholm 13-15 June 1997* (Stockholm Studies in Classical Archaeology, 12), ed. C. Scheffer, Stockholm 2001, 45-53.

Moore, M.B. & D. von Bothmer 1976
CVA USA, The Metropolitan Museum of Art IV. New York.

Murray, A.S. 1889
Archaic Etruscan Paintings from Caere, *JHS* 10, 243-252.

Nielsen, M. 1993
Cultural Orientations in Etruria in the Hellenistic Period: Greek Myths and Local Motifs on Volterran Urn Reliefs, *ActaHyp* 5, 319-357.

Oleson, J.P. 1975
Greek Myth and Etruscan Imagery in the Tomb of the Bulls at Tarquinia, *AJA* 79, 189-200.

Pottier, E. 1922
CVA France, Louvre I. Paris.

Prayon, F. 1977
Todesdämonen und die Troilossage in der frühetruskischen Kunst, *RM* 84, 181-197.

Raab, I. 1972
Zu den Darstellungen des Parisurteils in der griechischen Kunst. Frankfurt-Bern.

Rastrelli, A. 1981
CVA Italia, Chiusi, Museo Archeologico Nazionale I. Roma.

Roncalli, F. 1965
Le lastre dipinte da Cerveteri. Firenze.

Schauenburg, K. 1970
Zu griechischen Mythen in der etruskischen Kunst, *JdI* 85, 46-81.

Schauenburg, K. 1973
Parisurteil und Nessosabenteuer auf attischen Vasen hocharchaischer Zeit, *Aachen Kunstbl.* 44, 22-29.

Simon, E. 1973
Die Tomba dei Tori und der etruskische Apollonkult, *JdI* 88, 27-42.

Simon, E. 1996
Schriften zur etruskischen und italischen Kunst und Religion. Stuttgart.

Sparkes, B.A. 1996
The Red and the Black. London – New York.

Thiersch, H. 1899
"Tyrrhenische" Amphoren. Leipzig.

Tiverios, M.A. 1976
Hoi Tyrrhenikoi (attikoi) amphoreis. He skese tous me pontiakous (etrouskikous) kai ton Nikosthene, *EphemArch*, 44-57.

Touchefeu-Meynier, O. 1994
Polyxene, *LIMC* VII, 431-435.

Tuna-Nörling, Y. 1997
Attic black-figure export to the East: The "Tyrrhenian Group" in Ionia, *Athenian Potters and Painters – the Conference Proceedings*, eds. J.H. Oakley, W. Coulson, O. Palagia, Oxford 1997, 435-446.

Walters, H.B. 1898
On some black-figured vases recently acquired by the British Museum, *JHS* 18, 283-301.

Vermeule, E. & S. Chapman 1971
A Protoattic Human Sacrifice?, *AJA* 75, 285-293.

Margit von Mehren
Department of Archaeology and Ethnology
Vandkunsten 5
DK-1467 Copenhagen K

THE IMPORT OF ATTIC POTTERY TO LOCRI EPIZEPHYRII. A CASE OF REINTERPRETATION

TORBEN MELANDER

Two red-figure fragments (**Plates 1-2**) preserved in the Collection of Greek and Roman Antiquities in the Thorvaldsen Museum in Copenhagen, and attributed to the Altamura Painter and the Niobid Painter respectively,[1] represent the point of departure for the ideas presented below.

As is the case for most of the items from Antiquity collected by the sculptor Bertel Thorvaldsen (1770-1844), information concerning where the fragments were found and how they were acquired is wanting. All we know is that they were in Thorvaldsen's possession when he returned to Copenhagen in 1838, after spending over forty years (1797-1838) in Rome.[2] Once in Copenhagen, the sherds were placed, together with the rest of Thorvaldsen's gift of his own works and his collections, in the museum that the artist had made a condition of his gift to the city. The museum opened in 1848, four years after Thorvaldsen's death.

Fortunately, as also applies to certain other pieces in Thorvaldsen's collection of Greek and Roman Antiquities, it has been possible via other channels to gain information on where the two sherds were found. As early as 1925, Beazley identified the fragment by the Altamura Painter as belonging to the same vase as fragments in the Collection of Antiquities in Schloss Erbach, Odenwald in Germany.[3] Some years ago, the present author realised that Thorvaldsen's fragment by the Niobid Painter was related to others in the same collection (**Fig. 1**).[4] The manuscript catalogue from 1808 in Schloss Erbach contains the information that both the Altamura Painter sherds, as well as those by the Niobid Painter, were found in Locri. In documents relating to the collection, but subsequently destroyed by fire, it was further noted that the sherds were bought by Count Franz during a visit to Naples. They were acquired through the good offices of his personal friend, Ridolfi-

Fig. 1. Neck fragment from an Attic red-figure volute krater. Fleeing females. The Niobid Painter. Height 11 cm. From Locri Epizephyrii. Schloss Erbach, Erbach, Odenwald.

no Venuti, who is believed to have obtained them from his brother Niccolò Marcello Venuti.[5]

It is known that the sherds by the Altamura Painter in Erbach and the one in Copenhagen were drawn by M.W. Tischbein for the fifth, unpublished volume of Sir William Hamilton's second collection of vases, *Collection of Engravings from Ancient Vases Mostly of Pure Greek Workmanship... Now in the Possession of Sir William Hamilton Vols. 1 – 4*, 1791-1795, either because they belonged to Sir William Hamilton at that time, or because they were known to a wider circle and considered so valuable that they had to be included in the planned fifth volume.[6]

The fragments by the Altamura Painter thus seem to come from some of the earliest known excavations at Locri which, according to surviving accounts, started in the 1780s. The sherds by the Niobid Painter probably also derive from the same early excavations, although it cannot be excluded that Thorvaldsen's fragment was found separately during later excavations. In either case, this would still have taken place before Thorvaldsen left Italy[7] and, as such, they would still belong to "the old excavations", undertaken long before P. Orsi started more systematic investigations of Locri at the beginning of the 20th century.[8]

Accepting that the fragments were found at Locri, an even more precise location may be suggested, based on circumstantial evidence. At least two bronzes in Thorvaldsen's collection of Greek and Roman Antiquities can provide, if not proof, then at least some indication of where the find was made. One of the two bronzes is the handle of a vessel representing two lions attacking a hind or antelope.[9] In 1981, W. Gauer published the lower part of a lion's hind legs with the attachment plate in the Museo Archeologico in Reggio Calabria. According to the information available, the piece was found during "the old excavations" in the Persephone sanctuary, below the fortified Manella ridge in the north-eastern part of Locri (in this case "the old excavations" means the excavations by P. Orsi). At the same time, W. Gauer confirmed the near certainty of the bronze fragment's relationship to the handle portraying attacking lions in the Thorvaldsen Museum (or possibly the handle's counterpart). Consequently this must also have been found in the same locality at Locri.[10]

The other bronze in Thorvaldsen's collection of Greek and Roman Antiquities is a bull protome (**Plate 3**) which has an exact counterpart in the Museo Archeologico in Calabria. This piece is also known to have been found during "the old excavations" of the Locrian Persephone sanctuary.[11]

Although the two Thorvaldsen bronzes, and their now more accurately defined discovery site at Locri, do not prove that the sherds by the Altamura Painter and the Niobid Painter were found at exactly the same place, their presence in the same collection seems, nevertheless, to point to a common discovery site for the entire group, prior to their distribution among numerous owners.

In addition, the apparently highly fragmented remains of the two red-figure kraters also suggest a shrine, rather than a grave, as a likely finding place. Among Locri's many sanctuaries, the Persephone sanctuary[12] would be a strong probability. An examination of the subjects of the two krater fragments in the Thorvaldsen Museum and Schloss Erbach will, I hope, confirm this link.[13]

On the Altamura Painter sherd in the Thorvaldsen Museum, the remains of three figures are preserved (**Plate 1**). To the far left there is an outstretched, open left hand. Then comes part of the shoulder and head (looking back) of a woman walking towards the right. She is dressed in a chiton and himation, and her hair is tied by a scarf with a spotted ribbon. In front of the woman, we see the chest and head of Poseidon who is also dressed in a chiton and himation. In his right hand he holds the trident, in the other hand a small section of an *aphlaston* is left. Around his head he wears a wreath or ribbon embroidered with 'beads'.

On the large Erbach sherd only the uppermost curved part of a bow is preserved. The other Erbach fragment (a piece sawn from the sherd that was originally listed in the catalogue in 1808) shows a head of Athena wearing a diadem, set with two horse protomes, and an ornamental earring.

As far as it is possible to interpret from the preserved figures, we have the impression of a procession of gods in which Poseidon and Athena are immediately identifiable. The others could be Hermes (the outstretched hand), Amphitrite (immediately behind Poseidon) and Artemis or Apollo, possibly Herakles (the bow on the large Erbach sherd). Considering the krater's surface space all the way round, there is easily room for the whole host of Olympian gods on some errand or other.[14]

The sherd by the Niobid painter in the Thorvaldsen Museum (**Plate 2**) represents a woman running to the left towards a volute altar with a palm behind it. She wears a chiton and himation, and her hair is arranged at the back in a *krobylos* fixed with a diadem. On the far right there are the remnants of a female figure moving towards the right; at the top, part of a *krobylos*, at the bottom, a foot and the train of a chiton.

On the left of the Erbach sherd (**Fig. 1**), which shows the rest of the woman on the right of the Thorvaldsen sherd, a woman is seen running to the right. On the right-hand side of the fragment, a woman is running to the left, looking backwards (her face and left arm are missing). Both women are dressed in chiton and himation. The woman on the left has a tainia around her hair; the woman on the right has loose, flowing hair at the back and is wearing a hair band. This, insofar as it can still be seen, is more like the diadem on the Thorvaldsen sherd than the tainia worn by the other woman on the Erbach sherd. At the feet of the woman on the right on the Erbach sherd there is a kalathos. On both the Thorvaldsen and Erbach sherds the women's gestures express dismay, which is appropriate for a group of women running "in all directions".

Fig. 2. Locri tablet showing Persephone being abducted by a youth. Terracotta. After Prüchner 1968.

Together, the two sherds display the remains of a scene well known from the Niobid Painter and other vase painters: a group of women busy with female occupations (the kalathos) in the courtyard of the house (altar and palm tree) have been scared by a male intruder. If we compare this with other "pursuit scenes", by the Niobid painter, it is not impossible to recognise the scene as yet another depiction of Peleus forcing his way into Neleus' house in order to abduct Thetis.[15]

The scenes on our two sets of fragments showing, on the one, a "procession of the gods" and, on the other the abduction of a bride – or at least a man's interruption of the girls' domestic idyll can be interpreted in a context specific to their discovery site at Locri.

Here the so-called Locri-tablets come into the picture.[16] In particular, two main groups are of interest. The first involves abduction scenes: a young man abducting a young girl and taking her to a waiting chariot (**Fig. 2**), sometimes with the more or less apparent use of force, at other times, seemingly, without any force at all. Or the young man might be

Fig. 3. Fragment of a Locri tablet, showing Persephone being abducted by Hades. Terracotta. After Prüchner 1968.

climbing into the chariot holding the girl in his arms. Sometimes, the girl is seen with a kalathos in her arms; sometimes the kalathos has fallen from her. Far rarer are the scenes when the otherwise expected Hades drives off with the girl (**Fig. 3**). The second main group shows Persephone and Hades, enthroned, awaiting the congratulatory procession of gods or ordinary votaries (**Fig. 4**). Tablets depicting an indi-

Fig. 4. Fragment of a Locri tablet, showing Hades and Persephone enthroned. Terracotta. After Prüchner 1968.

vidual god standing before the pair of enthroned gods are related to this second group; those showing an isolated deity have similarly been seen as depicting individual figures in such a procession of gods.

The description above summarises the commonly held view of the contents of the tablets. In 1973, C. Sourvinou-Inwood suggested[17] that most of the Locri tablets showing abduction scenes refer to the Locri cult of Aphrodite and, in particular, to the simulated, cult-conditioned, abduction of the bride that was performed in Locri as part of the marriage ritual.

It must be emphasised here that both C. Sourvinou-Inwood's interpretation of the contents of the "abduction tablets" and H. Prüchner's view[18] that some of the tablets found in the Persephone shrine had fallen from a sanctuary higher up on the Manella ridge (and, as such, were intra-mural in contrast to those of the Persephone sanctuary) are in harmony with the relationship shown above between the two series of Locri tablets and the imported Attic ceramics portraying the two corresponding motif cycles of abduction and procession. Consequently, in the present context the two views will not receive further attention.

The correlation of the two types of objects, the local Locri tablets and the Attic ceramics depicting the procession and abduction scenes, aims to use Locri as a case study. Specifically, it aims to demonstrate that the use made of some ceramics in the places importing them is the result of a reinterpretation of the myths and legends of Attic material to suit local concepts. We cannot answer whether the reinterpretation had already been made when the order was sent to Athens, or whether it occurred in some intermediate place or even locally, possibly on a more or less individual basis. Who undertook the selection in each individual case, whether it was a trader, someone responsible for the cult or a third party, are questions that, presumably, cannot be answered today; in any case they require further research. This knowledge is certainly necessary in order to understand the conditions for the reinterpretation at a local level, but is not necessary in deciding whether, in some cases, a reinterpretation took place at all.[19]

Table A presents the numbers of Attic black and red figure vases showing the following motifs: *women pursued or abducted by gods or men, women occupied in the home,* and *satyrs pursuing nymphs*. The vases originate from various selected localities. **Table B** shows the vase shapes from the same localities listed in table A. The choice of localities, apart from their relative geographical proximity to Locri, is based solely on their degree of representativity, that is to say the number of vases discovered at specific localities in Southern Italy and Sicily.

THE IMPORT OF ATTIC POTTERY TO LOCRI EPIZEPHYRII.

	Total	bf/rf	1. Woman pursued by a youth	2. Woman pursued by a man	3. Woman pursued by a god	4. Woman fleeing	Total 1-4 bf/rf	5. Woman at home	6. Satyr/nymph	Total of 1-6 bf/rf : bf+rf
Locri	252	19/233	0/6	0/4	0/2	3/16	3/28	0/32	0/24	3/84 : 87
Taranto	196	121/75	1/3	1/0	0/0	1/10	3/13	8/12	8/3	19/18 : 37
Camarina	107	4/103	1/1	0/2	0/0	0/1	1/4	0/36	0/7	1/47 : 48
Gela	511	99/412	0/5	0/0	0/12	0/3	0/20	3/77	10/17	13/114 : 127
Agrigento	141	42/99	0/0	0/3	0/2	0/2	0/7	0/13	5/10	5/30 : 35
Paestum	35	16/17	1/0	0/0	0/0	0/0	1/0	0/4	3/0	4/4 : 8
Capua	286	69/217	2/6	2/2	0/6	0/0	4/14	0/33	12/21	16/68 : 84
Total	1528	370/1158	5/21	3/11	0/22	4/32	12/86	11/207	38/82	61/365 : 426

Table A. Selected motifs on Attic black-figure / red-figure vases from different localities in South Italy and on Sicily

	Amphorae A/B	Neck-amphorae	Panathenaic Amphorae	Pelikai	Hydriai	Kraters, Column-	Kraters, Volute-	Kraters, Calyx-	Kraters, Bell-	Stamnoi	Psykters	Oinochoai	Olpai	Skyphoi	Cups	Lekythoi	Alabastra	Aryballoi	Pyxides	Lekanides	Rhyta	Head-Vases	Amph.-Type?	Krater-Type?	Other types	?	Total bf/rf
Locri	5/6	2/5	1/1	0/4	1/8	1/34	2/14	0/14	0/28	1/1		0/9		3/13	2/17	0/35	0/2			0/2	0/6	0/2	1/0	0/12	0/0	0/20	19/233
Taranto	7/0	4/3	3/0	0/3	0/5	0/4	2/1	0/6	0/3	0/1		0/5		17/4	70/6	17/24	0/1	0/2					0/1	0/1	1/2	0/3	121/75
Camarina	0/1	0/0	0/0	0/4	1/9	0/11	0/1	0/12	0/9	0/0				1/3	0/5	2/36	0/0				0/1			0/6	0/1	0/4	4/103
Gela	2/1	21/49	0/0	1/20	2/3	2/31	0/4	0/10	0/8	0/3		0/4		8/6	5/16	55/249	0/1	0/0	1/0	2/0		0/3	0/3	0/1	0/1	0/2	99/412
Agrigento	7/3	4/0	1/0	1/4	0/2	1/23	0/2	0/14	0/6	0/2	0/1	2/2		0/5	0/1	25/26	0/0				0/1	0/3		0/0	1/3	0/1	42/99
Paestum	1/2	1/0	0/0	0/1	0/0	0/1	0/0	0/0	0/1	0/0		1/0		0/1	4/0	8/9	0/0							0/0	1/0	0/1	16/17
Capua	5/2	11/43	3/1	3/21	1/9	1/5	4/23	0/1	0/10	0/11		10/8		1/10	22/35	2/1	1/0			0/2	0/16	0/13		0/0	8/0	1/2	69/217
Total	27/16	41/99	8/2	5/56	5/56	5/109	4/23	0/65	0/65	1/18	0/1	13/28		29/42	101/80	107/372	1/4	0/2	1/0	2/4	0/23	0/24	1/0	0/20	10/7	1/33	370/1158

Table B: Shapes of Attic black- and red-figure vases from the same localities as mentioned in table A

67

As in most other cases, this analysis is based on the material identified and published by J.D. Beazley in Attic Black-Figure Vase Painters (1956), Attic Red-Figure Vase Painters 2.ed. (1963), and Paralipomena (1982), with appropriate reservations concerning the representative quality of the material as it is used here. This applies in particular to the black-figure material. The discrepancy between the totals in the tables (the first column in table A and the last in table B) and the number of vases in Beazley's indexes denoting provenance derives from my corrections of minor errors in the indexes.

The present analysis does not take *procession scenes* into account. But if, for instance, the *pursuit motif* in the case of Locri should differ from the overall count for this motif, we could also wonder whether other motifs, including the *procession motif*, might also have been reinterpreted in accordance with the special ideas in Locri, even if this shows no statistical effect on the totals. On the other hand, a count has been carried out of vase paintings showing scenes of *women occupied in the home* – a motif that has its counterpart in the many Locri tablets. These show *a)* Persephone busy with clothes chests, or *b)* they show just the household effects, the clothes chest, kalathoi etc. in a kind of short hand version of this activity.[20]

Finally, the *satyr pursuing nymph* motif is included as a special category of the motif of pursuit. In particular, I have considered the motif's relation to cultic plays or some other performances in which the fertility cult makes use of lascivious gestures and other forms of obscenity in keeping with the lustful behaviour of the satyrs. It is well known that obscenity was an important element in the cult of Demeter. This is clearly brought out in the Demeter Hymn's reference to the servant girl Iambe/Baubo's *skommata*, in the form of erotic jokes and mischievous inventions.[21] At the risk of over-interpreting an element that might have been purely decorative, reference must be made in this context to the satyr-nymph motif appearing on some of the Locri tablets in connection with furniture etc.[22]

It must be said straight away that the count in **Table A** does not provide an unambiguous picture, and certainly not a significantly unambiguous one. But based on the sample available, it is completely satisfactory if the figures merely indicate the probability of *reinterpretation* in a given context. It goes without saying that the figures must be read against the background of the overall occurrence of black-figure and red-figure ceramics at a given locality. For instance, if we ignore the number of vases with scenes illustrating the motif *women occupied in the home,* the figures for Gela are not impressive when seen in relation to the large total number of vases found. The total from Gela is twice that from Locri, which ranks third in terms of the total numbers discovered. In percentage terms, Gela ought to have a far higher proportion of these particular motifs – if, the ceramics had been imported without reference to the content of the motif.

The number of vases decorated with the *pursuit* motif (total of columns 1-4 in **Table A**) appears to be the category showing the clearest trend. Furthermore, in Locri it is also the most striking; this can be read as indicating that the motif was more favoured here than in the area (Southern Italy/Sicily) in general. An explanation for this could lie with local cultic situation.

By comparison, the motif *women occupied in the home* (**Table A, 5**), is far more evenly distributed between the sites when the individual numbers are read with the total numbers in mind. So, only the apparently special position of the *pursuit* motif in Locri justifies the suggestion that *the woman occupied in the home* motif was also subjected to a local interpretation but without its standing out statistically in the numbers of finds. Elsewhere, a quite different meaning may have been ascribed to the motif. This can be seen to some extent from the figures in the last subject column in the table: *satyr pursues nymph*. Here Locri has the second highest number, but the Capua material exceeds it by almost a third. However, the distribution of vase shapes in the same localities, shown in **Table B**, might provide an explanation of this.

As to the distribution of vase shapes in selected localities (**Table B**), it can be noted that no really significant numbers can be established in any of the localities. Nevertheless some trends may be noted at the various centres allowing us to draw both positive and negative profiles. With its complete lack of olpai and psykters, apart from the few "chance" finds of the latter vase shape in the Agrigento material, the present count also shows that there were differences not only in imports to the individual localities, but also to the major cultural areas.

Perhaps this is best illustrated by the distribution of stamnoi that appear in quite small numbers, in some cases there are none at all, in the selected localities. The exception is Capua which, in this context, stands out partly on account of this very vase type. The distribution of the stamnos is virtually limited to Etruria and the Etruscan spheres of influence far down in Campania.[23] So their presence in considerable numbers in Capua should not surprise us.

The same circumstance might also explain the high frequency, mentioned above, of the *satyr pursues nymph* motif in Capua as compared to its occurrence in Locri; even though it is more often met here than in any of the remaining localities. If we take into account the import of cups and oinochoai, where the motif recurs particularly frequently, this explains Capua's greater share in the motif. Again, it must be the influence of Etruscan culture asserting itself. The large number of cups imported to Taranto is not of great significance in this context, as a large proportion of the Attic cups from Taranto are earlier, from a period when the motif had not yet found the prominent place in the decoration repertoire which it gained later.

The occurrence of the other vase shapes listed in **Table B** might also deserve a commentary, but let us concentrate on the finds from Locri. Here, the most characteristic type of vessel is quite definitely the krater. To the four columns with identified kraters must further be added the third to last column "Krater type?" and the unidentified last column. Taking their share of the overall total into account, some, if not most, of the altogether 20 pieces listed in this final column must be krater fragments, especially as the more easily identifiable cup and lekyth fragments are unlikely to be included in this number. Some of the other localities also show an "interest" in some of the krater types, but without threatening Locri's dominance in the area in any way. The sole exception is Camarina where the krater is well represented, especially when seen in relation to the amount of Attic ceramics imported to Camarina in general.

If the occurrence of the *pursuit motif* on the kraters from Locri and Camarina is compared, however, Camarina does not appear to have the share of this motif which would be expected if imports had been made to the various localities without reference to motifs.

All in all, the impression seems to be confirmed that imported ceramics bearing a relevant subject were considered a suitable votive offering. When reinterpreted, as in the case of Locri, these could express the same as the much more frequently recurring votive tablets.

The far smaller number of "vase votive gifts" naturally leads to the question of the relative value of the vases and tablets. Were the vases only given as votive offerings because they happened to be available; possibly taken from the home when there was no longer any use for them? Or do they represent a valuable offering? That the imported vases were more expensive earlier in their existence than the votive tablets is obvious enough, but was this also true at the time when they were used as votive gifts? As my knowledge of the ceramics from Locri's Persephone sanctuary is limited to the pottery actually exhibited in the Museo Archeologico in Reggio Calabria, I cannot know whether some of the pieces in the museum's possession show marks of wear or even signs of repair after use. But the exhibited material, as I remember it, looked as though it was supplied directly from the place where the pots were produced. As far as the argument goes, it could be maintained that the small number of votive vases could suggest that it was not usual to make use of private pottery in a votive context. Again, the small number of vases compared to the number of votive tablets in itself justifies the assumption that a votive offering of the size and quality of a krater was considered a valuable and festive object that actually brought honour to the person presenting the votive gift. If this is correct, then the price was not without significance.[24]

Neither in counting the vase motifs nor the vase shapes has any distinction been made between grave finds and finds made in sanctuaries. The only distinction made in this commentary was the observation that the Thorvaldsen and Schloss Erbach sherds probably derive from a sanctuary; a conclusion based on the fragmented nature of the remains (cf. above). An examination of the vases found in tombs and exhibited in the Museo Archeologico in Reggio Calabria, and in the Antiquarium in the village of Locri[25] (the latter is, incidentally, close to the Greek necropoleis east and west of the city fortifications down on the coastal plain) gives the impression that the choice of motifs may, at times, have been motivated by local concepts of the cult of Persephone, with further local variations in Locri.[26]

Two of the motifs discussed above, *pursuit* and *woman at home*, are also to be seen on vases deposited in graves but here they are generally depicted on skyphoi and lekythoi. However, there is nothing to indicate whether the two vase types and the two motifs were equally divided between the graves of men and women, or those of boys and girls. In one case, "Tomba 400" in the museum in Reggio, a two-wheeled, cast bronze chariot, (*biga*) was placed in a grave together with a skyphos bearing scenes portraying A) a youth holding two spears[27] and pursuing B) a woman fleeing past an altar.[28] In the exhibition the chariot is referred to as a toy and, thus, the grave must be assumed to be that of a small boy although other interpretations are also possible. Irrespective of whether it comes from a child's or an adult's grave, we are allowed, with Locri's cult of Persephone in mind, not only to see the bronze chariot as a toy but, more importantly, to see the two "images" as one: the pursuit on the vase continues with the journey in a *biga* towards Hades – an image of death conforming to local ideas.[29] At the same time, the small grave find may also confirm the suggestion presented here that the placement of large kraters with the kinds of decorations mentioned above in the Persephone sanctuary could be assigned a new meaning and relevance in the local context.

A last circumstance, which has not been touched on before now, is the question of dating the relics discussed here. So far, only the names of the vase painters have been mentioned. The production of Locri tablets appears to begin at some time in the early 5th century BC and continues until c. 450 BC. A small early series still bears traces of the Archaic style that in the following series are replaced by Early Classical stylistic features.[30] For a thorough analysis of the groups, it is sufficient here to refer to P. Zancani Montuoro's great work in elucidating the complex developments in the use of shapes for the various series.[31]

A very small number of imported Attic vases from the Late Archaic period in Locri are decorated with the *pursuit* motif.[32] After these few "strays", comes the "bulk" of vases with the same motifs decorated by Early Classical pot painters.[33] As a whole the vases with the motifs *pursuing* and their dating correspond very well to the dates given above for the Locri tablets.

In elaborating their motifs, the coroplasters in Locri did not use the Attic imports as their models. Rather, they related to the current understanding of the myth – albeit with the exception of the youth's presence instead of Hades. Theoretically, it is possible that the youth was introduced into the scene on the Locri tablets due to the influence of the depictions on the imported Attic vases of youths forcing their way into a young female idyll. However, it is more likely that the youth should be seen as a local hero[34] who has been introduced into the story of Persephone. Such a youth is familiar in the local tradition from the production of terracotta figures representing *a mature man lying on a couch together with a youth.*[35] These figures have been found in both sanctuary and sepulchral contexts.

Finally, a speculation occasioned by the head of a satyr in Thorvaldsen's Collection of Greek and Roman Antiquities (**Plate 4a-d**) from a now lost bronze statuette.[36] The relevance of the head in the present context naturally depends on whether it is possible to demonstrate its belonging to a Locrian context. The head, which is 2.4 cm high, has been broken off just above the shoulder and chest section. The tip of the nose is slightly flattened. The peg projecting from the lower side of the neck is modern. The bearded head is distinguished by the pointed horse's ears that are characteristic of satyrs. The cheeks, nostrils and eyebrows are indicated by conspicuous incisions. The eyes are surrounded by powerful beaded edges. The forelocks are arranged in two rows of curls, and the hair at the back of the head is gathered in a *krobylos*. There is a *tainia* round the hair at the crown, differentiated by fine wavy lines emerging at the top of the crown and continuing to the forelocks and *krobylos*, where a large, vertical drilled hole 0.36 cm in diameter can be seen. The pronounced artery on the left of the neck and the *krobylos*, which has been asymmetrically drawn to the same side, show that the satyr's head must have been both inclined and turned to the right.

P.J. Riis has recently assigned the head to one of his late Vulcian series of bronzes.[37] The parallels noted by Riis to his Vulci school are not without interest, but if we examine the little satyr's head in detail, it seems to me that other parallels suggest themselves. In profile, the Locri tablets' portrayals of Hades in particular are of interest; both those in the abduction scenes and in the scenes representing the enthroned divine couple Persephone and Hades (**Figs. 3 and 4**).[38] We see the same profile, the same outline of the head, similarity in every single detail of the face and hair – and, in general, the same sensitive expression. This final characteristic enables the satyr's head to fit in easily with the other fine bronzes in the Thorvaldsen Museum that, as I mentioned above, were found in the Persephone sanctuary at Locri.[39] The close similarity to the style of the Locri tablets and, in particular, to their depictions of Hades naturally leads one to attempt to relate the satyr's head to the contents of the Locri tablets, an attempt that starts out primarily from the hole in the *krobylos*. In all probability, it served to fix something.

The question then is what was fixed? Representations of dancing satyrs or satyrs heavily burdened by wine amphoras, kraters or wine skins belonging to a wine orgy worthy of satyrs are not a rare sight in painted vases from the end of the 6th and beginning of the 5th centuries BC. Nor is the motif unknown among bronze figurines.[40]

Another object of satyrs' desire, and of their urge to carry burdens, are nymphs. This motif, which was extremely popular in Attic vase painting, is illustrated here by a detail from a decoration on a black figure oinochoe in Thorvaldsen's collection of Antiquities (**Plate 5**).[41] Two satyrs in a *Knielauf* design are shown – each carrying a nymph. Terracotta and bronze figures of the motif are also known. Among the bronzes, one remembers especially the Campanian group in the Metropolitan Museum, New York, where both the (kneeling) satyr and the nymph being carried on his shoulder have been preserved.[42]

It appears to be possible to discern two, or perhaps three, similarly shaped bronze statuettes of young women riding side-saddle, with one arm raised in alarm. The two identified statuettes are in the Museum für Kunst und Gewerbe in Hamburg[43] and the Antikensammlungen in Munich.[44] The latter still retains a large hand resting on her knee. The hand that belongs to the "gentleman" carrying her is drilled through in a similar manner to the hole drilled in the *krobylos* of Thorvaldsen's satyr. A third, similar statuette of a girl is found in the Antiquarium in the Florence Museo Archeologico, although I only know it from a reproduction in Reinach's *Repertoire statuaire*. In this, as the result of a presumably recently invented *compositum mixtum*, the statuette has been placed in the hands of a Hellenistic(?) bronze statuette.[45]

Whether the little bronze head in the Thorvaldsen Museum can be listed as yet another bronze statuette showing this motif depends, of course, primarily on whether the hole through the *krobylos*, in combination with the tilting of the head, supports such an attribution. Further, whether any of the nymph statuettes discussed here have been designed such as to fit in with the way the Thorvaldsen satyr head has been drilled through. Regarding the latter question, I have been able to examine both the nymph statuette in Hamburg and the group in New York. With its completely smoothly formed seat (the existing perforation is presumably modern and intended for fixing it when on display) fixation by soldering is all that seems feasible for the statuette in Hamburg,[46] while the New York group seems to be cast in one piece.

As already mentioned, the hole drilled in the Munich statuette probably represents a parallel, as far as the perforation is concerned, but it is positioned in such a way that the peg must have been inserted through/into the shoulder of the satyr carrying her. It is, of course, impossible to determine how the possible third example in Florence was fixed merely on the basis of the reproduction in Reinach. As regards the manner in which something could be fixed to the satyr's head itself, the above-mentioned concave traces of grinding across the top of the *krobylos* seem rather to suggest an elongated object, for instance an amphora like that referred to in connection with the satyr statuette in Boston.[47] But in that case, such a strong fastening as the existing hole drilled in the *krobylos* of the Thorvaldsen satyr head would not have been necessary.

It is scarcely possible to go further at the moment. If the stylistic parallel with the depictions of Hades on the Locri tablets can be sustained (which in the Thorvaldsen/Schloss Erbach context makes the Persephone shrine in Locri a likely discovery site) it is perhaps sufficient for the little head to be seen as part of a satyr game, perhaps a game in which several figures were included. The parallels to this, in a Persephone context, are the portrayals already discussed of satyrs and nymphs on the Locri tablets, in subordinate but nevertheless significant contexts.[48] And as further confirmation of the relevance of the satyr motif, there is the frequent appearance of the motif of a *satyr pursuing a nymph* on the imported Attic ceramics from Locri, as can be seen from the totals in the relevant column (6) in **table A.**

NOTES

I would like to thank the Ny Carlsberg Foundation and the Novo Nordisk Foundation for providing financial support for repeated visits to Reggio Calabria, Locri and other localities referred to in this article; I am also grateful to the Danish San Cataldo Institution at Scala (Italy) for providing hospitality for studies for this article in 1993 and 1998, and for shorter visits on my way to and from Locri.

1. (A) The Altamura Painter sherd (bell krater), Inv. H 596: Müller 1847, 71, cat. 96; Beazley 1963, 592, 37; Beazley 1982, 129; Giudice 1989, 16, note 5, no. 19; Melander 1993, cat. 37; CVA Th. Mus. 1, 11-12. (B) The Niobid Painter sherd (volute krater), Inv. H 602: Müller 1847, 77-78, cat. 102; Melander 1993, 74, cat. 38; CVA Th. Mus. 1, 11-12.

2. Especially on the collection of vases see CVA Th. Mus. 1, 11-18. On the transport to Denmark, which in fact comprised several shipments between 1835 and 1844 and thus gives rise to theoretical possibility of later additions to the collection, see CVA Th. Mus. 1, 14-15.

3. For J.D. Beazley's early ascription of the Altamura Painter sherd cf. Beazley 1925, 335, 22.

4. For the Thorvaldsen Museum sherd, above note 1, B; for the Schloss Erbach sherd Beazley 1963, 599-600, 10. In this case their belonging together is confirmed by the so-called *Bruch an Bruch* joint. The reverse of the Erbach sherd is labelled with information about finding place: Lokri (I saw the Erbach fragment during my stay in Schloss Erbach in 1986). For a rim fragment possibly belonging to the same vase, also in Erbach, see Giudice 1989, 64 note 361. My sincere thanks to the Gräflich Erbach-Erbach und Wartenberg-Rothische Rentekammer for alle help during my two visits in Schloss Erbach. I also acknowledge my thanks for the permission to publish the photo of the Niobid Painter sherd reproduced fig. 1.

5. Anthes 1895, 341-43. On the collection in general Prüchner 1981.

6. Reinach, 1900, pls. 334-365 reproduces the Tischbein drawings for the planned 5th volume of M.W.Tischbein, *Collection of Engravings from Ancient Vases.... Now in the Possession of Sir William Hamilton*, Vols. I-IV, Napoli 1791-1795; the Altamura Painter sherd is reproduced on pl. 364 (Tischbein pl.111).

7. On the early excavations in Locri see RE XIII, 2 s. v. Lokroi c.1289ff and Arias 1991, 201-204; on Duc de Luynes' excavations in 1828 below the Manella ridge see Duc de Luynes 1830, 3-12, with particular reference to pot sherds (black-figure) 7-8.

8. P.Orsi's excavations in Lokri Epizephyrii: Orsi 1909, 319-326; Orsi 1911, 77-124.

9. Inv. H 2029: Müller 1847, 160, cat. 29.

10. Gauer 1981, 115, 124-129, 133; Sabbione 1996, 33 and 35.

11. The Thorvaldsen protome Inv. H 2131: Müller 1847, 168, cat. 131. During my visits to the Museo Archeologico, Reggio Calabria (see note 25 below) the protome counterpart was to be seen in the section containing the finds from the Persephone shrine. In the protome in the Thorvaldsen Museum there is still a remnant of the core; the colour is the same characteristic light brown known from the Locri tablets (also below note 39). – In the archaeological literature, another bronze in Thorvaldsen's Collection of Antiquities has been connected with Locri: the "running" bronze youth Inv. H 2017: Müller 1847, 159, cat. 17. First by Fuchs 1969, 93-95, then Gauer 1981, 128, note 49. The bronze is presumably an Etruscan work and must be dated to c. 300 BC, Melander 1993, 120. See also Robertson 1975, 212.

12. For identification of the shrine below the Manella ridge in the northern outskirts of Locri, see Grillo 1996, 43-45.

13. An inquiry of Professor F.Giudice as to whether there might be sherds from the same krater among the other Niobid Painter sherds from Locri has not yet led to a positive result.

14. Within the oeuvre of the Altamura Painter for a very similar scene see Beazley 1963, 589, 3bis.

15. J.D.Beazley, who did not know the Thorvaldsen sherd, described the scene on the Erbach sherd "woman running to right, woman fleeing to left, looking round" (1963, 599-600, 10). But compare here the Niobid Painter krater in Bologna (Beazley 1963, 599, 8) with a neck frieze representing the Peleus-Thetis scene with a section very similar to the Thorvaldsen-Erbach scene. Referring to Sourvinou-Inwood's interpretation of the palm tree in this connection as a symbol of virginity: "Artemis as protector of *parthenoi* and of their preparation for marriage and transition to womanhood" (1991, 101) gives the possibility that the altar/palm tree in a palace suggests a special girl's section of *gynaikonitis*.

16. For a quick overview of the motifs on the Locri tablets, see the material collected by Prüchner 1968. Almost all Locri tablets have been found in the Persephone shrine below the Manella ridge; a few were found in other Locri shrines, and a few in houses. A special production of "Locri tablets" is known from Locri's colonies Medma and Hipponion, and now also from Franca near Naxos, Sicily, Borelli 1996, 10.

17. Sourvinou-Inwood 1973, 12-21.

18. Prüchner 1968, 63. For an overlapping of the two deities see Mertens Horn 1997 (same author in *Antike Welt* 1997, 3, 230).

19. For earlier suggestions concerning reinterpretation of the imported Greek ceramics, see Spivey 1991, 143-144: "The point is that a funerary gloss can be put on the Panathenaics: Panathenaics may be perceived outside Athens as *suthina*, as things suitable for grave decoration." – Suggestions for reinterpretation in connection with the import of Greek ceramics naturally do not exclude the possibility of ceramic ware being imported to the same locality as the result of, for instance, a special order (Lezzy-Hafter 1997, 359-364); or a production due to special interest in the Etruscan market (de La Genière 1988, 164-167).

20. Prüchner 1968, *a*: pl. 4; *b*: pl. 31.

21. Hymn to Demeter vv. 202-204; Simon 1969, 99.

22. For Locri tablets with the satyr/nymph motif on household goods, see Prüchner 1968, pl. 4; Sourvinou-Inwood's (1991,178) corresponding interpretation of the Aphrodite tablet where a satyr is shown penetrating a fawn on an altar.

23. de La Genière 1988, 161-162.

24. For the prices of vases see Johnston 1979, 33-35; see also Johnston 1991, 224-228.

25. My visits to the Museo Archeologico, Reggio Calabria: 1986, 1992, 1994 and 1996; in the Antiquarium in the modern village of Locri: 1994 and 1996.

26. Should Sourvinou-Inwood's interpretation of the abduction motif as being related to the bridal abduction tradition in Locri (note 17) turn out to be correct, this does not negate the reinterpretation concerning graves that merely moves from the Persephone to the Aphrodite cult. But see her indication of an influence from the cult of Aphrodite in the Locri Persephone, Sourvinou-Inwood 1991, 178-180. Cf. here also Mertens-Horn 1997, 106.

27. C. Sourvinou-Inwood's distinction between the individual pursuer's weapons etc. with a view to distinguishing the individual pursuits' myths/legends (1991, 27-143, where her three articles on this theme are reprinted with additions) is of no significance in the present Locri reinterpretation.

28. Beazley 1963, 974, 25, the Lewis Painter, c. 450 BC.

29. In the exhibition in the Museo Archeologico, Reggio Calabria, I have seen yet another two-wheeled bronze chariot, Inv. 25307.

30. On the dating of the Locri tablets to the first half of the 5th century BC, Vlad Borelli and Sabbione 1996, 40.

31. For P. Zancani Montuoro's treatment of the Locri tablets, see the short bibliography in *Santuari della Magna Grecia* 1996, 302: Zancani Montuoro 1954, 1957 and 1961.

32. For instance the Theseus Painter (Beazley 1971, 258); The Acropolis 787 Group (Beazley 1963, 1638,3); The Berlin Painter ibid. 221, 215; The Syriskos Painter, 266, 88.

33. For instance (probably), The Painter of the Yale Oinochoe (Beazley 1963, 503); The Syracuse Painter (521,47); The Aischines Painter (711, 77).

34. Vlad Borelli 1996, 40.

35. Cerzoso 1996, 123-124.

36. Inv. H 2009: Müller 1847, 158, cat. 9.

37. Riis 1998, 91, fig. 97; 122.

38. For Locri tablets with representations of Hades, see Prüchner 1968, pl. 12 (abduction scenes); pl. 22 (enthroned together with Persephone).

39. See above p. 61 for the small bronze group with provenance in Locri. The bull protome with the core suggests local production. For local bronze production in Locri see Jantzen 1937, 3-26; Langlotz 1963, 28-29.

40. For vases with representations of satyrs carrying wine jars, see for instance CVA Th. Mus. 1, pl. 54, and 50. For examples in bronze art see Mitten and Doeringer 1967, 85, cat. 80: Greek/Campanian satyr figurine bearing an amphora on his back, Walters Art Gallery, Baltimore, Inv. 54.2291, with references to further parallels.

41. CVA Th. Mus. 1, pl. 35.

42. Inv. 12.229,5: Richter 1915, 42-43, cat. 61. I am most grateful to Dr. J.R. Mertens for helping me in the Metropolitan Museum, as well as for sending photos of the group in question. A similar group in Indiana University Art Museum, Bloomington IN, mentioned by Dr. Mertens, is otherwise unknown to me. For terracotta figurines with the same motif, see Schneider-Hermann 1968, 110-113. The same article considers the contents of the pieces.

43. Museum für Kunst und Gewerbe, Hamburg, Inv. 1958.38, Mitten & Doeringer 1967, cat. 165. Dr. Cornelia Ewigleben helped me most kindly during my study of the bronze in Museum für Kunst und Gewerbe. So my thanks to Dr. Ewigleben too.

44. Staatliche Antikensammlungen, Munich, Inv. Br. 80, ex Coll. Dodwell. Antke und Abendland VIII, 1959, 127-131.

45. Museo Archeologico, Florence. Reproduced in Reinach 1897 pl. 443, a *compositum mixtum* in which a winged youth (Hellenistic?) is shown holding an archaic female figurine with a serrated diadem sitting with outstretched hands.

46. For a general study relating to complicated statuette compositions, in which the individual elements are cast separately, see Lechtman & Steinberg 1970, 5- 35; see also Bol 1985, 81-82; 136-138. On joining by means of pins, which the satyr's head in the Thorvaldsen Collection presupposes, see Lechtman and Steinberg 1970, 9. Perhaps joining in a lead mass might explain the large diameter of the hole drilled in the *krobylos*.

47. See note 40.

48. See above, notes 21 and 22.

BIBLIOGRAPHY

Anthes, E. 1895
Ein attisches Vasenfragment in Erbach, *Bonner Jahrbücher* 96, 341-343.

Arias, P.E. 1991
Locri, in: G. Nenci & G. Vallet (eds), Bibliografia topografica della colonizzazione greca in Italia e nelle isole Tirreniche. Rome 1991, 201-204.

Beazley, J.D. 1925
Attische Vasenmaler des rotfigurigen Stils. Tübingen.

Beazley, J.D. 1963
Attic Red-Figure Vase Painters (2nd ed.). Oxford.

Beazley, J.D. 1971
Paralipomena. Oxford.

Beazley, J.D. 1982
Beazley Addenda (Eds. L.Burn & R.Glynn). Oxford.

Bol, P.C. 1985
Antike Bronzeteknik. Munich.

Cerzoso, M. 1996
Medma: le recenti acquisizioni. L'area sacra al Mattatoio, in: *Santuari della Magna Grecia* 1996, 122-124.

Duc de Luynes, H. 1830
Ruines de Locres, *Ann Inst* II, 3-12.

Fuchs, W. 1969
Die Skulptur der Griechen. Munich.

Gauer, W. 1981
Ein spätarchaischer Beckengriff mit Tierkampfgruppe, in: *X. Bericht über die Ausgrabungen in Olympia.* Berlin, 11-165.

Genière, de La, J. 1988
Images attiques et religiosité étrusque, in: J. Christiansen & T. Melander (eds.), *Proceedings of the 3rd Symposium on Ancient Greek and Related Pottery* (Copenhagen 1987), Copenhagen 1988, 161-169.

Giudice, F. 1989
Vasi e frammenti "Beazley" da Locri Epizefiri. Studi e materiali di archeologia greca. Catania.

Grillo, E. 1996
Le testemonianze architettoniche del Santuario di Persefone alla Manella, in: *Santuari della Magna Grecia* 1996, 43-44.

Jantzen, U. 1937
Bronzewerkstätten in Grossgriechenland und Sizilien. Berlin.

Johnston, A.W. 1979
Trademarks on Greek Vases. Warminster, Wiltshire.

Johnston, A.W. 1991
Greek vases in the marketplace, in: T.Rasmussen & N.Spivey (eds.), *Looking at Greek Vases,* Cambridge, 203-231.

Langlotz, E. 1963
Die Kunst der Westgriechen. Munich.

Lechtman, H. & A. Steinberg 1970
Bronze Joining. A Study in Ancient Technology, in: S. Doeringer, D. G. Mitten and A. Steinberg (eds.), *Art and Technology. A Symposium on Classical Bronzes.* Cambridge, Massachusetts, 5-35.

Lezzy-Hafter, A. 1997
Offerings Made to Measure: Two special Commissions by the Eretria Painter for Apollonia Pontica in: J.H. Oakley et al. (eds.), *Athenian Potters and Painters. The Conference Proceedings,* Oxford, 353-369.

Melander, T. 1993
Thorvaldsens Antikker. Copenhagen.

Mertens-Horn, M. 1997
Rappresentazione di scene sacre, in: M.B. di Pralormo (ed.), Il Trono di Ludovisi e il Trono di Boston. Venice 1997, 94-118.

Mitten, D.G. & S.F. Doeringer 1967
Masterbronzes. Mainz.

Müller, L. 1847
Musée Thorvaldsen III, II. Copenhagen.

Orsi P. 1909
Lokroi Epizephyrioi. Quarta campagna di scavi (1909), *Nsc* 6, 319-326.

Orsi P. 1911
Lokroi Epizephyrioi. Quinta campagna di scavi (1910), *Nsc* 8, 1911 suppl. 67-75.

Prüchner, H. 1968
Die Lokrischen Reliefs. Beitrag zur Geschichte von Lokroi Epizephyroi. Mainz.

Prüchner, H. 1981
Die Römer Zimmer des Schlosses Erbach in Odenwald, in: H. Bech et al. (eds.), *Antikensammlungen im 18. Jahrhundert,* Berlin 1981, 237-255.

RE = Paulys Realencyclopädie der classischen Altertumswissenschaft. Stuttgart.

Reinach, S. 1897
Repertoire de la statuaire grecque et romaine I. Paris.

Reinach, S. 1900
Répertoire des vases peints grecs et étrusques II. Paris.

Richter, G.M.A. 1915
Greek, Etruscan and Roman Bronzes. The Metropolitan Museum of Art. New York.

Riis, P.J. 1998
Vulcentia Vetustiora. A Study of Archaic Vulcian Bronzes. Copenhagen.

Robertson, M. 1975
A History of Greek Art. Cambridge.

Sabbione, C. 1996
Il santuario di Persefone in contrada Manella, in: *Santuari della Magna Grecia* 1996, 32-39.

Santuari della Magna Grecia 1996
E.Lattanzi et al. (eds.), *Santuari della Magna Grecia in Calabria. I Greci in Occidente.* Napoli.

Schneider-Hermann, G. 1968
Silen mit Göttin und Nymphe, *AntK 11*, 110-113.

Simon, E. 1969
Die Götter der Griechen. Munich.

Sourvinou-Inwood, C. 1973
The Young Abductor of the Locrian Pinakes, *BICS* 20, 12-21.

Sourvinou-Inwood, C. 1991
'*Reading Greek Culture*'. Oxford.

Spivey, N. 1991
Greek Vases in Etruria, in: T.Rasmussen & N. Spivey (eds.), *Looking at Greek Vases*, Cambridge 1991, 131-150.

Vlad Borelli, L. & C. Sabbione 1996
I pinakes locresi della Mannella, in: *Santuari della Magna Grecia* 1996, 40-42.

Torben Melander
The Thorvaldsen Museum
Posthusgade 2
DK-1213 Copenhagen K.

GENUCILIA
SMALL PLATE WITH A LARGE RANGE

BIRTE POULSEN

Prompted by the forthcoming publication of a number of fragments of Genucilia plates found in the excavations of the Temple of Castor and Pollux in the Forum Romanum some years ago, I would like to touch upon some major questions associated with this group of pottery. These questions, which concern the purpose and function of the Genucilia plates, will be surveyed by investigating the plates' provenance and inscriptions: is it possible to relate the plates to any specific god/goddess or cult? Do they have a specific funerary function, or are they simply to be grouped together with other types of pottery meant for daily use? And, are these questions essential to the determination of the centre of production for these plates?

The Genucilia plate[1] is one of the most common ceramic forms of Mid-Republican Italy. This small and sometimes rather humble-looking plate is also among the most widely distributed groups of pottery in the Mediterranean area in the late 4th and early 3rd century BC. Both form and decoration are conveniently easy to determine. With the exception of only a very few plates,[2] the medallion shows two main types of decoration. On the majority of the plates the medallions are decorated with a female head in profile, facing to the left. On the basis of the half-sakkos (*sphendone*) worn by the women represented, Del Chiaro distinguished two different types of Genucilia plate.[3] The first he defined as the Falisco-Caeretan and Caeretan branch, in which the sakkos is normally cross-hatched, with the exception of a variant with a "star" pattern; the second, as the Faliscan type, in which the sakkos is usually decorated with a palmette. The woman's head is, however, frequently substituted by abstract or geometric motifs, in the Caeretan branch usually by a stylised star with four rays arranged round a central dot and with various kinds of motifs like dotted rosette, three dots arranged in a triangle or chevron placed in the quadrants between the rays (Del

Chiaro 1957, 283-287). The abstract designs of the Faliscan branch are more varied; they include such motifs as the rosette, quatrefoil and cross, with various "fillers"—small circles, dots, wriggling lines, chevrons—placed in the quadrants (Del Chiaro 1957, 288-298). This rather uniform decoration in the medallion of the shallow bowl is almost invariably encircled by a flaring rim with a wave pattern.

Provenance other than funerary
The existing distribution maps show that the Genucilia plates were widely exported. But it is remarkable that specimens of the two types almost never occur in the same sites (Jolivet 1982, figs. 72-73; Pianu 1985, figs. 5, 9-10). The Caeretan Genucilia plates, moreover, have been found in a far larger number of sites than the Faliscan ones, and also have a far wider geographical distribution.[4] Only a very few Faliscan specimens have been found outside the *ager Faliscus*, e.g. Ardea,[5] Naples or its vicinity (two Genucilia plates, both of Faliscan type, Jolivet 1980, 703, nos. 36-37), and Locri (see below). However, the main bulk of the Faliscan production clearly has a rather localized distribution. Although the total quantity of preserved Genucilia plates is high, and although we have a general impression of their distribution, it should be pointed out that the exact provenance is known for only a very few specimens found other than in funerary contexts. To ascertain what their possible function might have been, the following list (alphabetical) will include only non-funerary findspots.

Alba Fucens: One fragmentary plate found during the excavation of the sanctuary of Hercules (Balty 1969, 87, no. 1; pl. LVIa, fig. 24.1).
Artena: Three plates found in a *pozzo* of an unidentified building complex in the town area (L'édifice aux "thymiateria"); ten plates found in a *pozzo* near an unidentified building ("L'édifice à la citerne aux Genucilia") in the town area. All Caeretan (Lambrechts 1983, 88; *id.* 1989, 47-63).
Cori: Eight Genucilia plates found in a votive deposit associated with the sanctuary of Hercules, all Caeretan (*Enea* 1981, 29-34, A. 59-A62).
Caere: Three plates found in the area of Vigna Parrocchiale. They have *dipinti*, the letters *HPA*, written on the rim between the stylised waves. The three plates are all of the Caeretan type with a geometric motif in

the medallion. The excavated area was first interpreted as a sanctuary for Hera, but later Herakles has been suggested.[6]

Cosa: One rim fragment of a plate (Caeretan?) found in the backfill material of the Capitol (Taylor 1957, 79, pls. I and XXI; *Roma* 1973, 366, no. 533).

Elba: The hilltop fortress of Monte Castello di Procchio. Several plates of the Caeretan type with stars have been found. One of the plates has the inscription *"spurinies"*, painted before firing, below the base (Cristofani & Proietti 1982, figs. 2-4; on the inscription, Cristofani 1983, 214-215, no. 18, pl. XXVIII). Several fragmentary pieces were found during the first excavation campaign (Maggiani 1979, 10-15, nos. 9, 17, 22 and 59; Jolivet 1980, 717; Maggiani 1981, 185, pl. XLIXd). In a subsequent excavation, a pile of Genucilia plates were apparently found *in situ* in a cellar room interpreted as a food deposit (one of the plates is illustrated in Pancrazzi 1996, 34-35, 41, fig. 10).[7]

Ficana: At least nine fragments of Genucilia plates found in the habitation area, all Caeretan (Pietelä-Castrén 1999, 100-106).

La Giostra: Five fragmented plates found in the filling material. Presumably Caeretan (Moltesen & Brandt 1994, 92-93).

Lavinium: Several fragments of Genucilia plates, all Caeretan, found during the excavations of the sanctuary of Minerva (*Enea* 1981, 202-203, D 88).

Locri: Eleven plates, apparently found in a votive deposit in the sanctuary of Persephone (Fosse di Persefone). Eight plates with a woman's head, three with an abstract motif, all Faliscan (Del Chiaro 1957, 299-300; Jolivet 1980, 904, nos. 38-48; Pietilä-Castren 1999, 107 with note 68).

Ostia: *c.* 40 plates found in the filling material during the excavations of the Castrum, all Caeretan (Calza *et al.* 1953, pl. XXII, 3; *Roma* 1973, 355, no. 516; Lauro 1979).

Rome: From Rome itself Genucilia plates have been found in many Mid-Republican contexts, and it is significant that they all seem to belong to the so-called Caeretan branch. A large number derive from funerary contexts, but a small number were yielded by sanctuaries and votive deposits.

Excavations in three different places in the Forum Boarium have yielded several fragments: Three Genucilia plates were found near the fountain in the Piazza Bocca della verità (*Roma* 1973, 107, nos. 92-94);

other fragments were found during excavations in the Southern part of the Forum Boarium (*Roma* 1973, 105, no. 91; Lyngby & Sartorio 1968, 22 and 27, fig. 19.1); and finely two plates were recovered during excavations near the round temple (Lungotevere) (*Roma* 1973, 112, nos. 110-111). From the Palatine at least two Genucilia plates are known (Vaglieri 1907, 204, figs. 25-26; Romanelli 1951; Tomei 1997, 39, no. 21), but their exact findspots are unknown.

The largest number of the Genucilia plates from Rome come, however, from the Forum Romanum. Excavations in the Regia thus yielded eight Genucilia plates, three of which represent the well-known type with a woman's head in profile. A fourth specimen has a rather unusual motif, a representation of a ship, which seems to allude to the myth of Dionysos and the Tyrrhenian pirates. The remaining four plates have geometric motifs in the medallions;[8] they were found together in a *pozzo*, presumably a votive deposit. All the plates from the Roman Forum, including also, it seems, the one with the mythological representation, can be classified as Caeretan in type.

The known Genucilia plates from the Forum Romanum may now be supplemented by the excavated material from the Temple of Castor and Pollux: a total of 39 fragments: 27 rims, one base and 11 fragments of medallions. All except five fragments were found in stratified layers predating the Metellan reconstruction in 117 BC (Nielsen 1992, 30-41). None of the fragments are joining, but two may belong to the same plate. Part of the medallion is preserved only on ten fragments: five include remains of a woman's head (**Plate 6-7a & b**); the rest are decorated with a geometric motif (**Plate 8a-b**).

Unfortunately, part of the diagnostic headgear has been preserved in only one specimen, part of a cross-hatched sakkos, i.e. the so-called Caeretan type (**Plate 6a**). Of the remaining four fragments one retains the lower part of a woman's head with nose, lips, chin and neck with dotted necklace (**Plate 6b**); the style seems to be related to that of the so-called Carthage Genucilia Painter, an exponent of the Caeretan type (e.g. Del Chiaro 1957, 255-257). Another specimen preserves only the tip of the nose (**Plate 6c**), whereas a fourth shows the lower part of an ear with a curl in front of the ear with an earring (**Plate 6d**). The fifth fragment of a medallion is still attached to the only preserved base (**Plate 7a & b**, cf. below). Unfortunately, only part of the woman's front hair and outer part of eyebrow and eye have been preserved, but the drawing seems rather careless.

The five fragments of Genucilia plates with geometric motifs also seem to correspond to the Caeretan type. The geometric motifs of the Caeretan type may be classified into three groups according to the filling ornament in the quadrants between the rays (Del Chiaro 1957, 283-287, distinguishes three different types, 1-3). Only two of the preserved specimens retain enough of the medallion for a closer determination. One medallion has a dotted rosette between the rays (**Plate 8a**) like type 1, and the other (**Plate 8b**) a striped chevron, like type 3. One of the remaining three specimens apparently had no base, a feature which is not unusual for Genucilia decorated with stars (Del Chiaro 1957, 283).

The canonical decoration on the convex rim, consisting of stylised waves moving from right to left, is found on all the plates, with the exception of one single specimen (**Plate 8c**), which instead has a cross-hatched decoration. The excavation of the Temple of Castor and Pollux has further yielded rim fragments of three black-glazed Genucilia plates; one of the fragments is entirely black-glazed without decoration, whereas the other two are embellished with a laurel wreath in superimposed white (**Plate 8d-e**).

Satricum: One plate with star, Caeretan type, found in a votive deposit near the Temple of Mater Matuta (Della Seta 1918, 294; Beazley 1947, 104; *Roma* 1973, 336, no. 486).

Tor Tignosa (Pomezia): Sanctuary, the so-called "heroon of Aeneas" (Guarducci 1971, 88, Taf. 62.2-3).

Determination according to inscriptions (dipinti/graffiti)

It is well known that the group owes its name to a *dipinto* (*P. Genucilia*: Beazley 1947, 10), and in fact several of the plates are characterized by such *dipinti* and *graffiti* (Del Chiaro 1957, 293-296; Cristofani 1985). The only preserved base among the Genucilia plates from the Temple of Castor and Pollux has no less than three inscriptions on its underside, all apparently incised *after* firing. The simpler *graffito* consists of three short obliquely placed lines (*///*) near the edge of the base. The two remaining inscriptions evidently consist of Latin letters, of different length and style. Of the shorter word only the upper part of two letters is preserved: two vertical lines, *I* or *L*, followed by possibly the upper part of an *A*. The longer inscription contains six letters carefully incised along the rounded edge of the base. The first three letters are complete, *MAT*. The lower part of the remaining three letters is missing. But the fourth letter is clearly an *R*, while the fifth consists of a vertical line, perhaps an *I* or an *L*. The last letter resembles an *Y*, but is completed by a cross bar on top, presumably a wrongly written *T*. Assuming this to be the case, the word would then read *MATRIT*. The word yields no meaning directly, but if the word is divided it could be construed as *MATRI*, a dative of Mater, and a single *T*,[9] i.e a dedication to one of the Mater goddesses.[10]

In view of the generally accepted date of the production of the Genucilia plates only one of the Mater goddesses may be considered, namely Mater Matuta, since the cult of Magna Mater was not introduced into Rome until 205/204 BC (Latte 1960, 258-263). On the other hand, Mater Matuta was one of the old deities of Rome (Link 1930; Latte 1960, 97; Castagnoli 1979: Heldring 1985, 71-72). Originally she was a goddess not only of dawn, but also of birth and navigation. Her feast, the Matralia, was celebrated on 11th June, the date of the consecration of her temple in the Forum Boarium in Rome. The typical votive offerings to Mater Matuta were small cakes, *testuacium*, which according to ancient rites were baked in earthen forms (Varro, *Ling.* 5.106). The final *T* of the inscription could perhaps refer to these cakes (*testuacium*); perhaps they were presented to the goddess on these very plates. It was the custom on the feast of Mater Matuta for women to pray for their nephews and nieces. The feast was partly celebrated near the Temple of Mater Matuta, partly in private houses, and consequently this feast included both the public and private spheres (Beard & North & Price 1998, 50-51). Such an interpretation would seem to be confirmed by the fact that Genucilia plates were found in association with the temples of Mater Matuta in the S. Omobono area and in Satricum. The fact that these rites were also performed in the private sphere would explain why Genucilia plates have in some cases also been found in habitation quarters.

A small group of Genucilia plates with *dipinti* proves, however, that they cannot have been made exclusively for the cult of Mater Matuta (cf. Cristofani 1985). Three specimens found in the area of Vigna Parrocchiale in Caere have the letters *HPA* written on the rim between the stylised waves. The three plates are all of the Caeretan type with a geometric motif in the medallion. On the basis of these plates with *dipinti* the excavator concluded that the sanctuary was dedicated to Hera, and for many years other scholars accepted this theory (cf. note 6). Some years ago, however, the late M. Cristofani suggested that these *dipinti* should rather be interpreted as abbreviations for Herakles, like the *H* and *HP* found on other types of pottery. Similar abbreviations are well known from other pottery groups in Rome, namely the black-glazed cups with the Latin letter *H*, *HV* and *HVI* painted under the base. These cups were primarily found during the excavation of the foundations for a Mithraeum near the Circus Maximus, an area in which Hercules is known to have had several sanctuaries. The letters have thus been interpreted as abbreviations for *Herculi*, *Herculi Victori*, and *H(erculi) V(ictori) I(nvicto)* (Pietrangeli 1940, 144). So the three Greek letters *HPA* should perhaps be interpreted as an abbreviation for Herakles rather than Hera, an interpretation that may be supported by a fragment of a terracotta statuette representing Herakles found in the same area in Caere (Cristofani 1985, 24). The association of the Genucilia group with the cult of Hercules/Herakles is certainly confirmed by the relatively large number of plates found in connection with sanctuaries of Hercules, such as the specimens found by the excavations of the Lungotevere in Rome, in the sanctuary in Alba Fucens and the sanctuary in Cori.

Funerary contexts
A rapid survey of the provenances demonstrates, however, that the majority of the preserved Genucilia plates come not from votive deposits but from funerary contexts, as testified both in Rome and in other findspots further afield. Thus the necropolis in Aleria produced a surprisingly large number of Genucilia plates, a total of 118 distributed in 22 tombs (Jehasse & Jehasse 1973, 86-87). This distribution was not uniform: the Genucilia plates were entirely absent from many tombs (e.g. Jehasse & Jehasse 1973, 82 out of 105 tombs), while large concentrations were found in others, e.g. in tomb 33 where as many as 27 Genucilia plates were found in the dromos (Jehasse & Jehasse 1973, nos. 401-427), and in tomb 53 which yielded 16 plates (Jehasse & Jehasse 1973, nos. 801-816). In most cases, however, only one or two plates were extant among the grave-goods of the tombs in Aleria (Jehasse & Jehasse 1973, e.g. tomb 10, no. 11, and tomb 97, nos. 2084-2085). This

small number is also the norm when Genucilia plates are found in other sites further afield, like the tombs in Carthage and Sardinia.[11]

The same picture emerges from other necropoleis close to the presumed production sites, e.g. in Caere where some of the tombs contain a large number of Genucilia plates (e.g. tomb 54 of the Bufolareccia necropolis, Cristofani 1987, 319-320, no. 157) and Falerii, where a tomb contained as many as 12 Genucilia plates (Cozza & Pasqui 1887, 314). That the usual number is only one or two specimens is, however, demonstrated by the late tombs in Tarquinia in which the Genucilia plates are represented.[12] A tomb in Ardea contained two Genucilia plates as the only grave-goods (Pasqui 1900, 59). In Rome eight Genucilia plates were deposited in different tombs on the Esquiline,[13] whereas a tomb near the Circonvallazione Cornelia in the suburban area of Rome produced a total of seven Genucilia plates related to three burials (*Roma* 1973, 249-258, nos. 380-386).

The same tendency may also be observed among the Genucilia plates found in the necropoleis in Genoa (Jolivet 1980, 700, no. 27), Todi,[14] and Veii (Inglieri 1930, 56, fig. 7a). The specimens found in Populonia may also stem from a funerary context (Milani 1905, 56, fig. 2; Minto 1934, 417, fig. 71.6; Martelli 1981, 421; Bachielli 1986, 377, note 4; Bruni 1992, 68, fig. 42).

Purpose and function

A survey of findspots shows that the exact provenance of Genucilia plates found in non-funerary contexts is, unfortunately, in many cases uncertain.[15] Most significant are the Genucilia plates found during the excavations in Elba in a cellar room, which has been interpreted as a food deposit, and this context seems to indicate a profane or daily use. A similar purpose may be indicated by the specimens from habitation areas like the ones found in Ficana, Ostia, and perhaps Artena.

The known votive deposits are indeed few and do not connect the Genucilia with any particular deity: the plates have been found in conjunction with sanctuaries of Persephone (Locri), Mater Matuta (Rome and Satricum), Hercules (Caere?, Alba Fucens, Cori and Rome), Minerva (Lavinium), Hera (Caere?), Castor and Pollux (Rome), and the cults of the Regia (Rome). The recent finds from the Temple of Castor and Pollux, all found in the pre-Metellan fill, could perhaps be related to earlier construction phases of the temple as evinced by several of the other finds. It is a well-known fact that older votive deposits are buried in or near the sanctuary to which they were dedicated. Such an interpretation seems contradicted, however, by a *graffito* written below the only preserved base of a plate, assuming it really is meant as a dedication to Mater Matuta (**Plate 7 b**). The surviving *dipinti* do reveal, however, that at least some of the plates were made for specific gods or cult purposes.

Although the known number of Genucilia plates has increased considerably since Del Chiaro published his pioneering study, we are no closer to the solution of the purpose and function of this small plate (Del Chiaro 1957, 295-296). The form with a shallow basin and convex rim makes its use for drinking most improbable. The plates must rather have been intended for serving (or offering) small portions of dry goods such as dried fruits, nuts or small cakes (*testuacium*).

On the basis of a survey of the known provenances we may conclude that they served both as votive offerings, as funerary objects (Pietilä-Castren 1999, 107), and perhaps as plates for daily use. We may assume that they were made for some kind of profane or cultic purpose, and that they often ended up among the grave-goods in tombs; the same also holds good for many other groups of contemporary pottery. A comparative analysis of both the grave-goods from the tombs and the votive deposits shows, moreover, that the Genucilia plates with representation of a woman's head in the medallion occur together with the ones with geometric motifs. We may therefore deduce that the woman's head was not apparently considered of decisive importance in relation to the cults or to the burials. Furthermore, no specific iconographic interpretation can be applied to the appearance of the woman's head, e.g. as a representation of Mater Matuta or another goddess,[16] since the type frequently occurs also in other groups of Etruscan red-figured pottery (e.g. Del Chiaro 1974, 63-86). The head of the woman may presumably be regarded as a reduced representation of the female bust, often growing from acanthus scrolls, and so very common in the various groups of South Italian pottery (Trendall 1989, e.g. figs. 6, 179, 182, 186-187, 190-191, 210, 227, 254, 274, 290).

The question of production centres
In 1957 Del Chiaro argued (326-329), that there existed two main centres for the production of the Genucilia plates: Falerii Veteres and

Caere. The Faliscan production has never been questioned, but the existence of the Caeretan has been the subject of much discussion. Whereas the main bulk of the specimens identified as Faliscan derives from the *ager Faliscus*, only about half of the Caeretan Genucilia plates included by Del Chiaro in his survey had actually been found in Caere or the *ager Caeretanus*. As shown by the distribution maps (Jolivet 1982, figs. 72-73; Pianu 1985, figs. 5, 9-10), the specimens characterized as Caeretan are found in many different sites and were widely distributed.

The presence of an incised Latin alphabet on the rim of a Genucilia plate from the *ager Caeretanus* and the name in Latin letters below the base of another, both incised in the clay before firing,[17] have led some scholars to suggest that the so-called Caeretan production should rather be located in Rome.[18] One further argument in favour of Rome as a production site has been that some of the plates found in Rome are distinguished by some peculiarities in their representations (Jolivet 1982, 19, n. 128; Jolivet 1985, 58-69 with ref.). On the other hand, the name of a certain "Spurinies" made below the base of a Genucilia plate of the Caeretan type made *before* the firing (cf. Cristofani & Proietti 1982, figs. 2-4) surely indicates that at least part of the production was situated in Caere; moreover Genucilia plates classified as Caeretan do show affinities with other fabrics produced in Caere, and they are frequently found together with other widely exported fabrics from this very same site.[19]

Although the relatively large number of Genucilia fragments recently found during the excavations of the Temple of Castor and Pollux in the Forum Romanum does provide us with new information about the number of plates found in Rome, the question of the centre of production may still not be solved. The identifiable representations on the preserved fragments of Genucilia plate from the Temple of Castor and Pollux do seem to belong to the Caeretan group, a fact in accordance with the previously published finds from Rome. The only preserved base among the material has a Latin inscription engraved on the underside after firing; it cannot therefore be considered evidence of the production centre. It may be noted, however, that the clay and the fabric of the fragmented plates found in the Temple of Castor and Pollux are very uniform and could easily stem from one and the same workshop. The same goes for the glaze, although minor variations in colour do occur. With the exception of the rim fragment with cross-hatched decoration (**Plate 8c**) instead of the standard pattern of stylised waves, the decorative motifs also seem to conform to the decoration of the other specimens so far known from Rome. The arguments against Genucilia plates having been produced in Rome are not least based on the fact that there is no evidence for the existence of an independent production of red-figured pottery in Rome (for arguments for maintaining Caere as a production site, Cristofani & Proietti 1982; cf. Jolivet 1985, 58-59). The question may perhaps be answered through scientific analyses such as magnetic intensity or thermoluminescence sensitivity in future studies of the group, but until further evidence is available, it seems prudent to maintain Caere and Falerii Veteres as the main centres of production.

* Warm thanks are due to Marjatta Nielsen who offered constructive criticism of the manuscript, and to Peter Spring and Robin Lorcsh Wildfang who revised my English.

NOTES

1. These plates were first classified as an isolated group of Etruscan red figure by J.D. Beazley in 1947, 175-177, 303. Later the Genucilia plates were treated in detail by M.A. Del Chiaro in several studies, the first in 1957, which included a corpus of 600 specimens. Since then the published number of Genucilia plates has increased considerably, e.g. the material from Artena (Lambrechts 1983 and 1989), the Hellenistic tombs in Tarquinia (Cavagnaro Vanoni & Serra Ridgway 1989; Cavagnaro Vanoni 1996), the necropolis in Aleria (Jehasse & Jehasse 1973), and the collections in the British Museum (Bacchielli 1976), and in the Louvre (Jolivet 1984).

2. Plate found in the Regia with the representation of a ship, *Roma* 1973, 380, no. 558, pls. 24 and 26; Del Chiaro 1974, 66, fig. 5; Jolivet 1985, 64, fig. 5. Plate from Populonia with *ketos*, Martelli Cristofani 1979, 325, pl. 58. Plate from Cerveteri with three male heads in outline between the rays of a star, Del Chiaro 1974, 67, fig. 6. One of the Genucilia plates from Lavinium, now on exhibition in the Museo delle Navi at Nemi, displays a male head (unpublished).

3. The classification attempted by Del Chiaro 1957, 246-250, still holds good in broad outline, cf. the review by Colonna 1959. For discussion of the term *sakkos*, Del Chiaro 1974, 63.

4. Del Chiaro 1957, 297-305; Pianu 1980, 119-123; Jolivet 1980; Jolivet 1982, figs. 72-73; Bacchielli 1986, 375, with note 15.

5. Six Genucilia plates, all Caeretan, Pasqui 1900, 56 (tomb b), 57 (tomb e), 59 (tomb p). One Faliscan plate was found at Ardea, Del Chiaro 1957, 269-270, no. 1, pl. 21a, with ref.

6. Excavated in 1913, Mengarelli 1936. This identification was also accepted by Del Chiaro 1957, 295, who proposed that the Genucilia formed part of a votive offering to Hera. The identification as a sanctuary for Hercules was suggested by Cristofani 1985.

7. I thank Marjatta Nielsen for these informations.

8. *Roma* 1973, 380-381, nos. 558-592, pls. XXIV-XXVI; Del Chiaro 1974, 66-67, fig. 5; Cristofani & Proietti 1982, 72; Jolivet 1985, 63-64.

9. Such abbreviations are known from other classes of pottery in Rome, e.g. *HVI* standing for *H(erculi) V(ictori) I(nvicto)*, Pietrangeli 1940, 144. Cf. Cristofani 1985 for similar abbreviations in the Genucilia group.

10. Dedications to different gods are predominant on another group of pottery from this period, the so-called *pocola*, see *Roma* 1973, 57-67. I have been unable, however, to find any direct parallel for the combination of god/goddess and abbreviation for the votive offering. Early dedications to Mater Matuta are abundant on stone cippi, on which the name of the goddess appears in the dative and sometimes also the name of the dedicator in the nominative, Degrassi 1957, nos. 17 and 24. For these references I thank M. Skafte Jensen. For the discussion and interpretation of the inscription, I am indebted to O. S. Due, M. Skafte Jensen, and R. Lorsch Wildfang.

11. Carthage: two Genucilia plates in the Bord-el-Djedid necropolis, presumably decorated by the same painter, the Carthage Genucilia Painter (Del Chiaro 1957, 255-256, pls. 18-19; Morel 1980, 67-68, fig. 54): A third was found in the Sainte Monique necropolis (Ferron 1966, 703-704, pl. 25). Sardinia: a total of four plates; at least one was found in the Tuvixeddu necropolis, the rest perhaps in the habitation areas (Pianu 1985, 75-76).

12. Cavagnaro Vanoni & Ridgway 1989, tomb nos. 752, 1577, 4941, 5024, 5040, 5046, 5049, 5051, 5430, 5434, 5512, 5654; Cavagnaro Vanoni 1996, nos. 752, 1577, 5430, 5434, 5436, 5512, 5654.

13. *Roma* 1973, 209, no. 284; 211-213, nos. 292, 294-296; Scott Ryberg 1940, 56, 58-59, 86, 88, 101-102, 106, pl. 20, 107a-c. A further three specimens are entirely black-glazed, *Roma* 1973, 212, no. 293, 213, no. 297, and 229, no. 351.

14. One plate with a woman's head may clearly be attributed to the Caeretan group. The other two plates also have women's heads, but they are rather different, having long hair at the nape of the neck, *CVA Todi* 1940, pl. 12. 10-12.

15. E.g. Cumae: two Genucilia plates, both Caeretan (Jolivet 1980, 702-703, nos. 34-35), Palestrina: one plate, Caeretan? (Del Chiaro 1957, 257), Ampurias: one plate, Caeretan (Jolivet 1980, 686, no. 3), Cyrene: one plate, Caeretan (Bacchielli 1976), Malta: one plate, Caeretan? (MacIntosh Turfa 1977, 371, no. 97), Marseilles: one plate (Del Chiaro 1957, 263, no. 5), Olbia: two plates (Jolivet 1980, 697, note 12).

16. Only after finishing this paper my attention was drawn to the article by Pietilä-Castren 1999. She argues that the Genucilia plates are going to be seen as implements of the cult of Persephone in the area around Rome during the 4th century BC.

17. From Monteroni di Palo: Gasperini 1972-1973; Cristofani & Proietti 1982, 70; Civiltà degli etruschi 1985, 343-344, no. 3, with previous literature. Jolivet 1985, 64, argues that the letters have certain features of Etruscan palaeography and may have been incised after firing.

18. *Roma* 1973, 45 and 52; Torelli 1981, 80 with note 38; Rome had already been suggested as a production site by Scott Ryberg 1940, 101-102 (Falerii or vicinity and perhaps Rome); Beazley 1947, 10 (Rome, some other Latin city or Etruria). Del Chiaro 1957, 301, does not exclude that Genucilia plates were also manufactured in Rome. Cf. also Jolivet 1985.

19. Pianu 1985, for distribution maps of Caeretan and Faliscan pottery. The political relations between Caere and Rome during this period do not preclude the plates from having been made in Caere: Sordi 1960; Cristofani & Proietti 1982, 71-72; Bacchielli 1986, 375.

BIBLIOGRAPHY

Bacchielli, L. 1976
Un piattello "di genucilia". I rapporti di Cirene con l'Italia nella seconda metà del IV sec. a.C., *QuadALibia* 8, 99-107.

Bacchielli, L 1986
I piattelli Genucilia, in: J. Swaddling (ed.), *Iron Age Artefacts in the British Museum*, Oxford, 375-378.

Balty J. Ch. 1969
Observations nouvelles sur les portiques et le sacellum du sanctuaire Herculéen d'Alba Fucens, in: J. Mertens (ed.), *Alba Fucens II*, Bruxelles/Rome, 69-98.

Beard, M. & North, J & Price, S. 1998
Religions of Rome. Cambridge.

Beazley, J.D. 1947
Etruscan Vase-Painting, Oxford.

Bruni, S. 1992
Le ceramiche con decorazione sovradipinta, in: A. Romualdi (ed.), *Populonia in età ellenistica. I materiali delle necropoli (Atti del seminario Firenze 1986)*, Firenze, 58-109.

Calza, G. et al. 1953
Scavi di Ostia I, Topografia generale. Roma.

Castagnoli, F. 1979
Il culto della Mater Matuta e della Fortuna nel Foro Boario, *StRom* 27, 145-152.

Cavagnaro Vanoni L. & Serra Ridgway, F.R. 1989
Vasi etruschi a figure rosse dagli scavi della Fondazione Lerici nella necropoli dei Monterozzi a Tarquinia. Roma.

Cavagnaro Vanoni, L. 1996
Tombe tarquiniesi di età ellenistica. Catalogo di ventisei tombe a camera scoperte dalla Fondazione Lerici *in Località Calvario*. Roma.

Civiltà degli etruschi 1985
In: M. Cristofani (ed.), *Civiltà degli etruschi*. Milano.

Colonna, G. 1959
Review of Del Chiaro 1957, *ArchCl* 11, 134-136.

Cozza, A. & Pasqui, A. 1887
Civita Castellana, *NSc*, 307-319.

Cristofani, M. (ed.) 1983
Rivista di epigrafia etrusca, *StEtr* 51, 1983, 195-314.

Cristofani, M. 1985
Altre novità sui Genucilia, in: *Contributi alla ceramica etrusca tardo-classica*, (*QuadAEI*) 10, 21-24.

Cristofani, M. 1987
La ceramica a figure rosse, in: M. Martelli (ed.), *La ceramica degli Etruschi. La pittura vascolare*. Novara.

Cristofani, M. & Proietti, G. 1982
Novità sui Genucilia, *Prospettiva* 31, 69-73.

CVA Todi 1940
CVA Musei Comunali Umbri (Museo Comunale di Todi). Roma.

Degrassi, A, 1957
Inscriptiones latinae liberae rei publicae I. Firenze.

Del Chiaro, M.A. 1957
The Genucilia Group. Berkeley/Los Angeles.

Del Chiaro, M.A. 1974
Etruscan Red-Figured Vase-Painting at Caere. Berkeley, Los Angeles & London.

Della Seta, A. 1918
Museo di Villa Giulia. Roma.

L'Elba 1979
A. Maggiani & O. Pancrazzi (eds.), *L'Elba preromana: fortezze d'altura. Primi resultati di scavo. Monte Castello di Procchio, Castiglione di S. Martino*, (Exh. cat., Portoferraio 1979), Pisa.

Enea 1981
Enea nel Lazio. Archeologia e mito, (Exh. cat.),Roma.

Ferron, J. 1966
Les relations de Carthage avec l'Etrurie, *Latomus* 25, 689-709.

Gasperini, L. 1972-1973
Alfabetico modello latino su piattello etrusco del "Gruppo di Genucilia", *AnnMacerata* 5-6, 527-537.

Guarducci, M. 1971
Enea e Vesta, *RM* 78, 73-118.

Heldring, B. 1985
Mater Matuta, la dea di Satricum, in: *Satricum*, Latina, 68-77.

Inglieri, R.V. 1930
Veio, *NSc*, 45-73.

Jehasse, J. & Jehasse, L. 1973
La nécropole préromaine d'Aleria, (Gallia suppl. 25). Paris.

Jolivet, V. 1980
Exportations étrusques tardives (IVe-IIIe siècles) en méditerranée occidentale, *MEFRA* 92, 681-724.

Jolivet, V. 1982
Recherches sur la céramique étrusque à figures rouges tardive du musée du Louvre. Paris.

Jolivet, V. 1984
CVA France 33, Louvre 22. Paris.

Jolivet, V. 1985
La céramique étrusque des IV-III s. à Rome, in: *Contributi alla ceramica etrusca tardoclassica, (QuadAEI* 10), 55-66.

Lambrechts, R. 1983
Artena I. L'édifice aux "thymiateria", Bruxelles/Rome, 73-94.

Lambrechts, R. 1989
Artena II. "L'édifice à la citerne aux Genucilia", Bruxelles/Rome, 13-100.

Latte, L. 1960
Römische Religionsgeschichte. München.

Lauro, M.G. 1979
Una classe di ceramiche ad Ostia: il guppo Genucilia, *RStLig* 45, 51-66.

Lyngby, H. & Sartorio, G. 1968
Indagini archeologiche nell'area dell'antica porta Trigemina, *BCom* 80, 1965-1967, 5-36.

Link, H.C. 1930
Matuta, in: *RE* XIV.2, Stuttgart , cols. 2326-2329.

Maggiani, A. 1979
Monte Castello di Procchio, in: *L'Elba* 1979, 5-29.

Maggiani, A. 1981
Nuove evidenze archeologiche nell'isola d'Elba, in: *L'Etruria Mineraria, Atti del XII Convegno di Studi Etruschi e Italici, Firenze-Populonia-Piombino 1979*, Firenze, 173-192.

Martelli, M. 1981
Scavo di edifici nella zona 'industriale' di Populonia, in: *L'Etruria Mineraria, Atti del XII Convegno di Studi Etruschi e Italici, Firenze-Populonia-Piombino 1979*, Firenze, 161-172.

Martelli Cristofani, M. 1979
Davvero tarquiniese la "Tarquinia Silhouette Workshop"?, *AnnAcEtr* 18, 319-327.

MacIntosh Turfa, J. 1977
Evidence for Etrusco-Punic relations, *AJA* 81, 368-374.

Mengarelli, R. 1936
Il luogo e i materiali del tempio di *HPA* a Caere, *StEtr* 10, 67-86.

Milani, L.A. 1905
Campiglia Marittima, *NSc,* 54-70.

Minto, A. 1934
Populonia, *NSc*, 351-428.

Moltesen, M. & Brandt, R. 1994
Excavations at La Giostra, (AnalRom suppl. 21). Rome.

Morel, J.P. 1980
Les céramiques à vernis noir et à figures rouges d'Afrique, *AntAfr* 15, 29-75.

Nielsen, I. 1992
General description of the ruin and the trenches, in: I. Nielsen & B. Poulsen (eds.), *The Temple of Castor and Pollux*, Roma, 30-41.

Pancrazzi, O. 1996
The first hall, in: *Museo Civico Archeologico Portoferraio*, Firenze, 33-35.

Pasqui 1900
Ardea, *NSc*, 53-69.

Pianu, G. 1980
Ceramiche etrusche a figure rosse. Materiali del Museo Archeologico Nazionale di Tarquinia. Roma.

Pianu, G. 1985
La diffusione della tarda ceramica a figure rosse: un problema storico-commerciale, in: *Contributi alla ceramica etrusca tardo-classica, (QuadAEI 10)*, 67-82.

Pietilä-Castrén, L. 1999
Genucilia plates – common *AGALMATA* or depictions of the myth of Persephone, *Arctos* 33, 93-110.

Pietrangeli, C. 1940
Il mitreo del Palazzo dei Musei di Roma, *BCom* 68, 143-173.

Roma 1973
Roma Medio-Republicana. Catalogo della mostra. Roma.

Romanelli, P. 1951
Gli strati paleorepubblicani, *MonAnt* XLI, 101-124

Scott Ryberg, I. 1940
An Archeological Record of Rome from the seventh to the second Century B.C. London/Philadelphia.

Sordi, M. 1960
I rapporti romano-ceriti e l'origine della civitas sine suffragio, Roma.

Taylor, D.M. 1957
Cosa: Black Glaze Pottery, *MemAmAc* 25, 67-193.

Tomei, M.A. 1997
Museo Palatino, Roma.

Torelli, M. 1981
Gli Etruschi e Roma, in: *Incontro di Studio in onore di M. Pallottino*, Roma, 71-82.

Trendall, A.D. 1989
Red Figure Vases of South Italy and Sicily. London.

Vaglieri, D. 1907
Roma, *NSc*, 183-211.

Birte Poulsen
Department of Classical Archaeology
University of Aarhus
DK-8000 Aarhus C

klabp@hum.au.dk

All photos: the author.

TERRACOTTA HOUSE MODELS FROM BASILICATA

HELLE DAMGAARD ANDERSEN &
HELLE W. HORSNÆS

Degne di considerazione mi sembrano anche alcune cassette cinerarie di terracotta, che fatte a guisa di piccole case si scambiano luce con quelle di particolar forma una volta scoperte ad Albano. Due di'esse sono dipinte, come se la casa fosse costruita di travi di legno riempite di opera muratoria, una terza a guisa di scacchiera. In cima del tetto tutti portano per ornato ossia acroterio varie teste d'animali domestici, come cavallo, asino, ariete, oppure uccelli. Ad introdurre le ceneri poi serve un'apertura quadrata nel tetto.

(H. Brunn, in *BdI* 1853, 163)

For over a century the Department of Near Eastern and Classical Antiquities in the Danish National Museum, Copenhagen, has owned a small terracotta model of a rectangular building. The model was acquired by the museum at the public sale of the Julien Gréaux collection on 16th August 1891. The sales catalogue of the collection gives Etruria as the provenance of the model, but the decoration of the model is wholly un-Etruscan (Frœner 1891, n. 59, pl. 1). In 1985 Buranelli (1985, 73) convincingly identified the model with one of the three models seen by Brunn in the collection of sig. Amati from Potenza and described in the citation above.

The present article will discuss the South Italian models more closely. The models will be dated by means of a stylistical analysis, and it will be suggested that a more precise provenance can be restored by means of this analysis in combination with an evaluation of the objects in the Amati collection as known through Brunn's description.[1]

Fig. 1. Copenhagen inv. no. 3732, side A (photo and drawing by Niels Levinsen).

Fig. 2. Copenhagen inv. no. 3732, side B (photo and drawing by Niels Levinsen).

1. The Copenhagen model
The Danish National Museum, Dept. of Near Eastern and Classical Antiquities, inv. no. 3732 (**Plate 9 and Figs. 1-4**).[2]
The model rests on four rectangular feet, carrying the corners of the building. Above each of the four legs a hole (diam. 4 mm) has been drilled through diagonally from the outside to a point of attachment between the legs and the bottom. The walls of the building are vertical, and on the four corners of each of the short sides is a knob-like projection (on side C the upper and lower projections on the left side are missing). The short ends terminate in triangular gables, and in both gables is a tall rectangular opening. The roof is two-faced and steep with an inclination of 57-66 degrees. The central panel of one of the falls of the roof is made up by a loose trap door. Along the ridge of the roof there are three, originally four, plastic birds with bulging eyes, attached into drilled holes. On either side of the eaves of the roof there are two similar holes, probably for attaching other terracotta ornaments or animals.[3]

The model is complete, but it has been restored from two large fragments, and the trap door has a number of breaks. The paint is well-preserved. As to the technique, the model is composed of clay plaques consisting of two short sides, two long sides, one bottom, and two sides of the roof, joined at the ridge. The legs are attached to the bottom and then cut at the edges. The knob-like projections on the short sides were attached separately. The two missing projections must have fallen off before the model was painted since the scars where they were attached can still be seen. The rectangular holes in the gables are cut out. The plastic handmade birds were attached separately into holes on the ridge (diam. 2 mm), probably by means of a wooden stick. The holes are placed 7.2 and 7.3 cm from the corners on side A, and 7.3 and 9.3 cm from the corners on side B. The loose trap door on the roof's A-side is cut out with diagonal edges in order to keep the door in place.

The model is covered by a painted decoration in red and black. The decoration on each side and on the falls of the roof is framed by a black band. The short sides are covered by a panel with a chessboard pattern in black and reserved (i.e. unpainted), framed by a red band, and the gable is red. Between the lower short side and the gable is a black line. The rectangular openings in the gables are framed by a black line. The long sides are divided into three panels. The central panel on side A consists

Fig. 3. Copenhagen inv. no. 3732, detail side A (photo and drawing by Niels Levinsen).

Fig. 4. Copenhagen inv. no. 3732, detail side B (photo and drawing by Niels Levinsen).

Fig. 5. Louvre, inv. no. MNB 473, long side (courtesy of Christian Larrieu, Musée du Louvre, Département des Antiquités Grecques, Étrusques et Romaines, photo by M. Chuzeville).

Fig. 6. Louvre, inv. no. MNB 473, short side (courtesy of Christian Larrieu, Musée du Louvre, Département des Antiquités Grecques, Étrusques et Romaines, photo by M. Chuzeville).

of a reserved rectangle, framed by a red band between two black lines. Within the rectangle are black dots neatly set in lines and columns. On each side of the rectangle there is another rectangle, consisting also of a thick red band. They are filled in by a net pattern in thin black lines. On side A the net pattern is irregular. A horizontal red band above the central panel joins the two side panels. On the roof the trap door on side A has a black/reserved chessboard pattern framed by a red band and a black line. The side panels have a red frame with a vertical bar hanging from the centre. Inside the frame there is a row of black dots. Attached to the hanging bar there is a black cock-like bird to the right. On the fall on side B there is a similar decoration, but here the central panel with a chessboard pattern is narrower, leaving more space for the two side panels with a bird decoration identical to the one on side A. The plastic birds on the ridge of the roof are painted brown. The feet are red on the exterior.

2. *The Louvre model*
The Louvre, Paris, inv. MNB 473.
A model now in the Louvre can be identified with another of the models described by Brunn (**Figs. 5-6**).[4] The model came to the Louvre in 1872, and was then said to have been found in Southern Italy (Pottier 1897, 36 cat. no. D 32). The measurements, form and technique of the Louvre model are similar to those of the Copenhagen model, but the Louvre model is slightly larger and differs in many details. Two horizontally-set holes have been drilled through all four legs. The short sides on the model are smooth, missing the projections seen on the Copenhagen model. At the centre of one of the long sides there is a circular, no doubt intentionally made, hole and there is a small hole at the centre of the trap door. The roof is not as steep as the roof on the Copenhagen model, c. 45 degrees. On the ridge of the roof there is a plastic decoration consisting of two birds in alignment with the ridge and on the top of each gable a ram's head. On each of the four lower corners are protomes resembling snakes' or birds' heads, and on the eaves there are three small triangular projections.[5]

The model is decorated with black and red paint. On the feet there are black horizontal lines. Each of the long sides has a panel with a chessboard pattern framed by a black line and a band of solid black triangles. The short sides, including the gables, have a rhombic "chessboard" pattern in black, and triangles and chevrons in black and red. The falls of the roof are decorated with a red zigzag band framed by two black lines.

3. A lost model?
The whereabouts of the third model mentioned by Brunn is no longer known. Brunn's description indicates that it was decorated in a manner similar to the Copenhagen model and had a plastic decoration on the roof, probably depicting the horse or donkey explicitly described by Brunn.

Stylistical analysis
The bichrome decoration on the Copenhagen and the Louvre models clearly indicates a connection to the South Italian Matt-Painted pottery (Yntema 1984). Within this area (**Fig. 7**) the shape of the two models is unique, and only a comparison with the decorative scheme of the painted pottery can provide a closer identification and dating of the models.

The closest parallels of the decorative patterns are found within the West-Lucanian Matt-Painted wares of the Subgeometric style (Yntema 1984, 186-191). The chessboard pattern prominent on both house models is introduced during the second stage of the Subgeometric group, c. 575-550 BC. The Subgeometric group can be subdivided into a simple style which Yntema argues was influenced by the Greek banded ware, and another, more richly decorated style, confined to a minor number of vase forms: the "Palinuro jar", the column krater and a single example of a *nestoris* (Yntema 1984, 188). Being covered entirely by decoration, the models are of course closely connected with the richly decorated group.

Yntema proposes that the stylistical difference between the two groups was due to different uses of the two types of pottery, the richly-decorated style having been made specifically for burials. This cannot be ruled out, of course, but one should remember that also the simple-style pottery has been found in tombs, and since other types of find contexts (domestic/cultic) are still extremely rare, it is today not possible to double-check this theory. Pushing Yntema's line of thought a bit further, the complex style may be seen as a conservative one, underlining the "non-Greekness" of these forms. This idea, however, fails to take into consideration the form of two of the vases: the *nestoris* is not

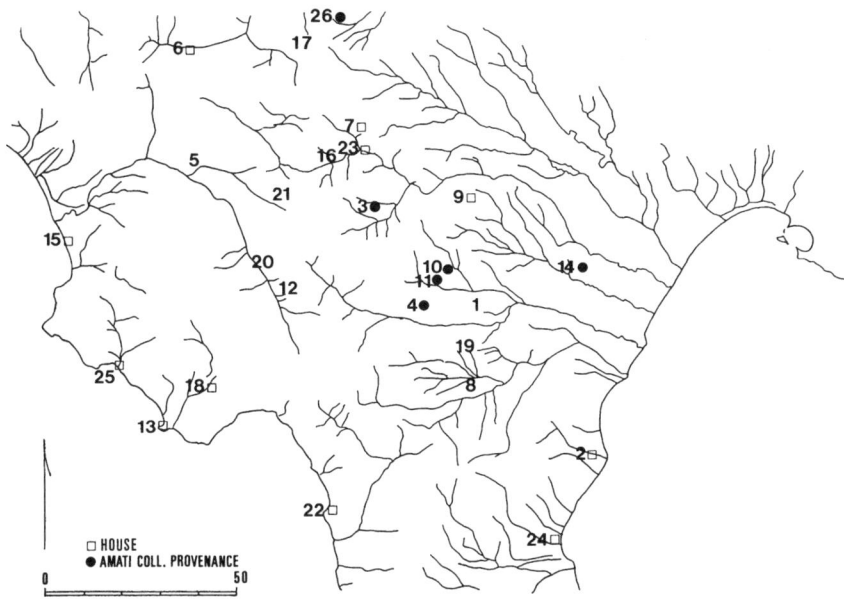

Fig. 7. Map of Basilicata and Southern Campania with sites mentioned in the text. Sites with remains of Archaic buildings are marked by a rectangle ☐, while sites mentioned as provenance for objects in the Amati collection are marked by a dot ●.

1	Aliano	14	Pisticci
2	Amendolara	15	Poseidonia
3	Anzi	16	Potenza
4	Armento	17	Ripacandida
5	Buccino	18	Roccagloriosa
6	Cairano	19	Roccanova
7	Cancellara	20	Sala Consilina
8	Chiaromonte	21	Satriano
9	Garaguso	22	Scalea
10	Gorgolione	23	Serra di Vaglio
11	Guardia Perticara	24	Sybaris
12	Padula	25	Velia
13	Palinuro	26	Venosa

a local form – it is found in only one example in Western Lucania – and the column krater is a purely Greek form, known in Matt-Painted ware in only four examples. We believe it is much more fruitful to see the complex style as one used on large vases – and on the house models – where it could be used freely without giving the overcrowded effect that could not be avoided on small vases.

At Garaguso in the Salandrella Valley, both the chessboard pattern and the red zigzag with a black outline are common, among others on kantharoi from tomb N, dated by the Ionic cups of type B2 from the tomb.[6] The solid triangles as well as the rhombic net pattern on the Louvre model have parallels on a group of "Oinotrian" kantharoi found at Garaguso (Nafissi 1985, 195). Furthermore, a similar rhombic net pattern is common on the body of the krater-kantharos form of Sala Consilina phases III C and early III D.

The broad frame lined with dots on the Copenhagen model is not found on pottery from the West-Lucanian group, but is instead encountered on vases from the North-Lucanian Ruvo-Satriano Class. However, as a whole the style of this class does not correspond as closely to that of the house model as does the West-Lucanian group (Bottini 1981).

In the publication of the first part of the Chiaromonte necropolis, Alfonsina Russo Tagliente (1992-1993, 276-277) has proposed that the Subgeometric "Oinotrian"/West-Lucanian Matt-Painted ware from the beginning was produced in different centres in the Vallo di Diano, the Val d'Agri and the Sinni Valley. The decoration of the "Oinotrian" pottery from Chiaromonte and Roccanova does indeed seem to be less complex than what is found on some of the vases from Sala Consilina (de La Genière 1968) and Palinuro (Naumann & Neutsch 1960). It should, however, be noted that the series from Roccanova ends around the mid-6th century BC (Tocco 1980). The apparent difference may therefore be due to chronological differences, or even – as argued above – to the use of different types of decorative schemes according to the size and form of the vessel.

The 5th century eaves tiles from Serra di Vaglio also present a bichrome decoration comparable to the house models: both red and black panels with chessboard pattern, and the red zigzag band with black outlines similar to the Louvre roof. The tiles may belong to the rectangular building A/78, dated to the mid-5th century BC by the rich pottery finds (Greco 1980, esp. 374-375).

The figured decoration
The silhouette birds on the Copenhagen model have no exact parallels. In general figured decoration is rare on "Oinotrian" pottery, but a

number of examples have been published from Sala Consilina. A krater-kantharos from tomb IV excavated in 1934 displays animals on both sides: on side A two quadrupeds to the right, and on side B two birds to the right. The krater-kantharos is dated to Sala Consilina phase III C.[7] A slightly later example comes from tomb A 52, dated in Sala Consilina phase III D, and a similar example is found on a fragment from Potenza (de La Genière 1968, 154, pls. 47.2, 47.5). An unusual frieze of deer to the right is seen on a kantharos from Garaguso (Nafissi 1985, 195, fig. 285).

This rarity of figured decoration is even more pronounced when it comes to human figures, as also noted by Tagliente in the publication of a newly recovered bowl on a stemmed foot, with a representation of a man surrounded by four birds on the interior. On the rim there is a chessboard pattern, and on the exterior alternating red and black bands. The bowl comes from the Santa Maria La Stelle necropolis of Aliano, but is unfortunately without context. It is dated in the late 6th century BC, presumably on stylistic criteria.[8]

Plastic decoration (birds) comparable to the roof decoration on the house models are found on two ring vases from Sala Consilina tombs B 58 and A 12, dated in late phase III C and early phase III D respectively (de La Genière 1968, pls. 46.2, 46.6), and on a flask from Padula, Valle Pupina tomb 46, also placed in the beginning of Sala Consilina phase III D.

The shape of the building models is fundamentally different from the anatomy of a vase. This difference gave the potter/artisan other possibilities for the decoration, and we should not expect a total accordance in the decorative schemes. The analysis of the decoration on the house models has revealed a number of *comparanda* among the "Oinotrian"/ West-Lucanian Matt-Painted pottery from the second half of the 6th (or perhaps even the beginning of the 5th) century BC and some likeness also among the North-Lucanian Ruvo-Satriano Class. The closest parallels are found among material dated in the 6th century, for example from Sala Consilina phase IIIC (often late) and IIID, and from Garaguso in contexts with Ionian type B2 cups. The stylistic analysis therefore suggests that the models were made in Western Lucania in the 6th century BC, probably in the second half of that century.[9]

Where were the house models found?
This stylistical identification is in accordance with the information given between the lines in Brunn's article on sig. Amati's collection. Most of the objects in the Amati collection came from Basilicata, and a number of sites are explicitly mentioned: Canosa di Puglia, Venosa, Anzi, Pisticci, Gorgolione, Guardia Perticara and Armento (**Fig. 7**). The objects provided with precise provenances are figured pottery, normally red-figured vases. Only from Guardia Perticara are recorded some black-figured lekythoi in addition to the red-figured vases – which probably were relatively early, as they are described as *severità nel disegno*.

The sites Anzi, Gorgolione, Guardia Perticara and Armento are all situated within the distribution area of the "Oinotrian" pottery from the 6th century to the early 5th century BC. Gorgolione is the only one of these sites that seems to have no "archaeological history" today.

At Anzi, excavations started already in the late 18th century. These discoveries comprised both tombs from the "Oinotrian" phase – vases with geometric decoration – and rich finds of Attic and Italiote red-figured vases (Pontrandolfo, in De Caro & Borriello 1996, 46-48; bibliography in *BTCGI* III 1984, 252-254).

The first finds from Armento were reported in 1814 and the site was intensively excavated in the first half of the 19th century. Few of the early finds can be identified today, but the descriptions leave little doubt that very rich tombs from the Lucanian period were excavated. At Armento, modern fieldwork may have identified the old excavation area in loc. Campo Scavo near Serra Lustrante, where a Lucanian sanctuary and habitation quarters have also been found. Material from the Archaic period is found at some distance, close to the modern village Armento, where a rich warrior's tomb from the beginning of the 6th century was uncovered in 1967.[10]

The excavations at Guardia Perticara that provided sig. Amati with objects seem not to be recorded. There can, however, be no doubt from the description of black-figured lekythoi and drawings in "severe style" of the red-figured pottery that tombs from the Late Archaic period had been uncovered. The recent excavations at Guardia Perticara have yielded tombs from the 7th century to the 4th century BC, but few details on the material have been published (Michelini, in *BTCGI* VIII 1990, 215-216 with references; Bottini & Barra Bagnasco 1990, esp. 556-557).

It is possible that the two house models came from one of the cemeteries in the area, where there is good evidence for 6th-century material: Alianello (Bianco *et al.* 1996, passim), Chiaromonte,[11] Guardia Perticara, or Roccanova. On these sites large numbers of tombs have been found, and the excellent state of preservation of both models would be explicable if they were found in tombs. The only site among these to be mentioned by Brunn as provenance for objects in the Amati collection is Guardia Perticara.

Other building models from Archaic Italy
In previous literature, the two models have been discussed only in connection with Etruscan and Latial hut and house urns. No doubt this was due not only to the original comparison with hut urns already made by Brunn, but also because the Drouot catalogue gave Etruria as the provenance for the Copenhagen model. Consequently, it has been exhibited among the Etruscan antiquities in the National Museum.

The models are usually compared to a series of funerary urns from Cerveteri, dated to the second half of the 7th century BC (Buranelli 1985; Coen 1991; Damgaard Andersen 2001). They have several features in common with the Copenhagen/Louvre models: all are rectangular, placed on rectangular feet, most of them have a painted decoration, all have a pitched roof, and several of them have plastic decorations on the roof. On the other hand there are many differences: the Copenhagen and Louvre models are much smaller and much narrower (on the Caeretan models the length varies from 30.5 to 55.5 cm; the width from 20.5 to 30.5 cm; and the height from 30 to 42 cm) and the roof is an integrated part of the models, whereas the roof-shaped lids on the Caeretan models are removable. The decoration on the Caeretan models consists of floral motifs as well as real and fantastic animals. Finally, as shown above, a stylistical analysis of the decoration clearly points to a South Italian production in the Archaic period.

A single building model was found in a tomb dated to 800-760 BC at Sala Consilina (Campania), but it was not used as an urn (**Fig. 8**).[12] The model shows a rectangular building. On the short front side there is a rectangular door, and on one of the long sides a small rectangular window. The roof is fairly steep and two-faced. At the top there is a volute, and behind each of these a plastic bird. The model is made of dark impasto and painted with geometric designs and zigzag lines on

Fig. 8. House model from Sala Consilina (Kilian 1962, fig. 19).

both the walls and the roof. On the windowless long side there is a meander, and on the rear short side there are three swastikas.[13]

A number of terracotta votive models have been found in sanctuaries in the Etrusco-Campanian area – Minturno, Teano, Capua, and Fratte – all dated to the 5th/4th century BC or later (Staccioli 1968, cat. nos. 41-58). These closely resemble Etruscan votive models of houses with tiled roofs, thus having no relevance to the Copenhagen/Louvre models.

Terracotta votive models are also known from Magna Graecia, especially from Locri and its sub-colonies (Danner 1992, 36-48). The Greek models differ from the Copenhagen/Louvre models in several respects: they depict a rectangular one-room building with a door on one of the short sides, a metope-triglyph frieze on the upper part of the walls, and a tiled roof with pan tiles and rows of cover tiles, as well as architectural terracottas. The finds from Locri can be dated to the late 6th/early 5th century BC. The finds from Medma were found in a deposit dated to the mid-6th/mid-5th century BC, but because of the triglyph in the gables, the models should probably be dated to the first half of the 5th century. The finds from Gorgoneia should probably be dated to the late 6th/early 5th century BC. The models from Hipponion cannot be dated precisely. The terracotta models can be compared to the famous building model in Greek marble from the votive deposit at Garaguso (Sestieri Bertarelli 1958).

The model as representation of architecture
The decoration on the long sides of the Copenhagen model clearly depicts a building with a light wall construction, probably wattle-and-daub or a light timber construction with *pisé* (rammed earth) or mudbrick. The upper horizontal band on either side and the knob-like projections, which are placed at the end of these bands, may be interpreted as tie beams and the ends of the wall plates, the lower horizontal band as the "foundation"/ground plate for the wattle-and-daub wall, the vertical bands representing the corner posts. The rectangle in the centre of one of the long sides may represent a door, and the hole in one of the gables a vent hole. Both the roof on the Copenhagen and the Louvre models are pitched, and the steep inclination (57-66 and 45 degrees, respectively) indicates that a thatched roof was intended.[14] The loose trap door is probably not a real feature of a house.[15]

The birds on the roof should be interpreted as wooden figures, cut out of the (not visible) ends of the cross-pieces keeping the thatched roof in place. Such plastic birds have many parallels, for example in the above-mentioned house model from Sala Consilina as well as in several Etruscan and Latial models. In addition, many Early Iron Age hut urns have bird-like animals cut out of the end of the cross-pieces (e.g. Bartoloni *et al.* 1987, fig. 4), as do one of the funerary urns from Cerveteri (Buranelli 1985, fig. 19). Finally, birds and human beings are seen on the roof of the huts in the carved decoration of the wooden throne from Verucchio, dated to the mid-7th century BC.[16] Birds are also known as acroteria, for example on the Archaic stone urns from Chiusi, dated to the 6th/early 5th century BC (e.g. Buranelli 1985, fig. 39).

The holes pierced through the bottom of the Copenhagen model and the legs of the Louvre models must have served a practical purpose. They may have been used to fasten the models to pedestals, for example to exhibit them in a manner similar to the almost contemporary house model from Caltanisetta (Schattner 1988).

Remains of buildings
Building remains from the 6th to 5th century BC are relatively rare in Basilicata. In her recent survey of domestic architecture, Russo Tagliente has collected much of the evidence.[17] The houses are rectangular, built of mud-brick on stone foundations, and with floors of beaten earth. The roofs may be made in either perishable material (probably thatched), or they were covered by roof tiles.

The rooms often measure approx. 5 x 5 m, and entrances and/or doors are rarely preserved. It is therefore difficult to determine whether two adjacent rooms are evidence of a two-room building, or whether they should be regarded as two separate houses sharing a wall.[18] Two cases of larger halls should, however, be mentioned: the apsidal hall at Cancellara, measuring 8 x 17.5 m (Russo Tagliente 1992, no. 10) and the well-known hall at Serra di Vaglio, loc. Braida (12 x 24 m), decorated with the terracotta friezes of Metapontine type and painted roof tiles. The building technique used for the halls does, however, not differ in technique from the houses.

The function of the building models
It is unlikely that the house models were votives, as were the Greek and Etrusco-Campanian models. Evidence of cultic assemblages in the "Oinotrian culture" in the 6th and 5th centuries is extremely rare, and in spite of the evident interaction between the Greeks and the indigenous peoples of Southern Italy in the Archaic period, it seems more appropriate to interpret the models on the premises of the indigenous culture.

The good state of preservation of the Copenhagen and Louvre models suggests that they were found in a tomb. The only other non-Greek model from Southern Italy, the well-known model from Sala Consilina, was not used as a cinerary urn, but placed in the tomb as part of the grave goods. Brunn's idea that the models were used as urns probably originated in analogy with the Central Italian urns. However, this is highly unlikely since the prevailing burial rite in Southern Italy is inhumation, and cremation is practically unknown in Western Basilicata in the Archaic period.[19]

The models could be considered symbolic representations of the house (of the deceased?) and thus a symbolic representation of the deceased himself (or herself), since there is a close connection between the house and the individual.[20]

In the 6th century, building models already had a long history in Italy, and the centres in the Agri and Sinni Valleys no doubt had relations both to Etruria and to the Greek cities along the Ionian Coast, as can be seen from imported luxury items found particularly in the princely tombs in these areas (Bianco *et al.* 1996, passim).

Still, the Copenhagen and Louvre models differ in many respects from the Etruscan and Greek models described here. It is therefore not possible to see them as close imitations of one or the other. Rather, they must be seen as indigenous objects and interpreted in light of the local cultures, the understanding of which will increase significantly when the results of the still on-going excavations in the Agri and Sinni Valleys become available for further studies.

Appendix – New results 1997-1999
During the 37th Convegno Internazionale di Studi sulla Magna Grecia (Taranto 1997), dott.ssa M.L. Nava, Soprintendente Archeologica di Basilicata, presented a "new" house model from a private collection in Basilicata. The model was clearly of the same type as the models discussed here, but the lack of a plastic decoration on the roof excluded the possibility that it should be identified with the lost model from sig. Amati's collection. Furthermore, dott.ssa Nava referred to another model, found in fragments in a 6th-century tomb during the 1997 excavations at Guardia Perticara. This model was characterized by the presence of plastic animals on the ridge of the roof, described by dott.ssa Nava as "dove-like", and thus comparable to the birds on the Copenhagen model.[21]

In 1998 a house model was included in the exhibition *"Tesori dell'Italia del Sud. Greci ed indigeni in Basilicata"*.[22] This model differs from the Copenhagen and Louvre models in two aspects: firstly, it rests directly on the ground, having no 'legs'; and secondly, it depicts a low-pitched roof that may represent a tiled rather than a thatched roof. The painted decoration of this model seems stylistically later than the previously known models, and thus suggests a date in the 5th century BC.

Another model, closely comparable to the Copenhagen and Louvre models, was illustrated in 1999 in connection with the exhibition *"Nel cuore dell'Enotria: la necropoli italica di Guardia Perticara"*.[23] It is interesting to note that this model has donkeys' heads as central acroteria, and it is thus closely reminiscent of the lost(?) third model described by Brunn.

The latest information implies that a total of five models have been found in female tombs in Guardia Perticara.[24]

The newly discovered models are of great importance, as they confirm the geographical and chronological conclusions based on the stylistical analysis of the Copenhagen and Louvre models.

NOTES

1. The present article was written in 1997, but circumstances have conspired to delay its publication. We are very grateful that the editors of the Acta Hyperborea have accepted the original version of the article for publication.
   Since the article was written, excavations taking place in Guardia Perticara in 1997 and 1998 have confirmed our hypothesis that the Copenhagen and Louvre models came from tombs in this area, cf. Appendix.

2. Bibl.: Brunn 1853, esp. 165; Frœner 1891, *Terres-cuites grecs*, n. 59, pl. 1; Strong 1968, 69 (ill.); Buranelli 1985, 73-76, fig. 41; Bartoloni, Beijer, De Santis 1985, 175-202, esp. 188, pl. 9,4; Colonna 1986, 393; Bartoloni et al. 1987, 142, tav. LVII,b; Stopponi 1985, 60, ad n. 2; Brijder, Beelen, van der Meer 1989, esp. 67 and 212, cat. no. 51 (ill.).
   Our warmest thanks to Dr. Bodil Bundgaard Rasmussen, Keeper of the Dept. of Near Eastern and Classical Antiquities, for permission to study and publish the model.

3. The models measures L. 25; W. 9.5; H. (to the ridgepole) 24.5 cm. It is made of light pinkish clay (Munsell 7.5 YR 7/4 (pink)) with a decoration in red (10 R 4/6), black (N 3/0), and brown (5 YR 4/2 to 3/3).

4. Bibl.: Brunn 1853, esp. 165; Pottier 1897, 36, cat. no. D 32, pl. 29; Pottier 1899, 367 and 392-93, cat. no. D 32, pl. 29; Pinza 1905, 618, fig. 189e; Richter 1966, 10, fig. 30; Buranelli 1985, 74-76, fig. 42; Colonna 1986, 393.
   We would like to thank Dr. Françoise Gaultier of the Louvre for information regarding this model.

5. The model measures L. 36; W. c. 15; H. 28.5 cm. The clay is light (Munsell 7.5 YR 7/3, pink).The model is intact, the plastic decoration is slightly restored.

6. Morel 1970, esp. 560 and fig. 26; see also tomb X, Hano, *et al*. 1971, esp. 430-431; Adamesteanu 1974, 132; Nafissi 1985, 195.

7. de La Genière 1968, pls. 14.1 and 15.2; Naumann & Neutsch 1960, Taf. 70.3; in both cases only the quadrupeds are illustrated.

8. Tagliente, in Bianco *et al.* 1996, 79-81 and 162, cat.no. 2.22, colour illustration on p. 83; earlier finds are discussed by Orlandini 1980.
Malnati 1984, esp. 50 sees these hourglass-formed figures as a typological continuation of the figures on 8th-century *a tenda* ware vases.

9. Since the publication of Yntema's study of matt-painted pottery, large cemeteries have been excavated in inland Basilicata: at Aliano, Chiaromonte, Guardia Perticara, Roccanova Serre etc., and these have yielded much additional material that – when published – may enable a revision of the proposed typology.

10. Adamesteanu 1970-1971. On the early excavations at Armento see Adamesteanu, in *BTCGI* III, 1984, 312-314; Pontrandolfo, in Bianco *et al.* 1996, 43; on the recent excavations in general see Russo Tagliente & Bianco, in Bianco *et al.* 1996, 238.

11. The relatively good state of preservation of the painting of the models renders it unlikely that they should come from Chiaromonte, Sotto La Croce necropolis, where the acidity of the soil leaves extremely few traces of the painted decoration (Russo Tagliente 1992-1993, 277-278; Bianco *et al.* 1996).

12. S. Antonio tomb 63 (*a pozzo*), excavated 10th September 1959. The bones and ashes were placed in a decorated biconical urn. In the tomb other types of impasto pottery were also found (two small *ollae* or jars, a bowl, and a bowl on a high foot), and a Sicilian fibula. The fibula indicates that the deceased must have been male. The model is now at Salerno, Museo Archeologico, inv. no. SC/14635.
Bibl.: Kilian 1962a, 64, 71, cat. no. 168, fig. 19; Kilian 1962b, 86-87, Abb. 2,11; 1969, 69, Taf. Va; Kilian 1970, 288; Edlund Gantz 1972, 189 and 192; d'Agostino 1974, 23, tav. 10; Buranelli 1985, 76; Colonna 1986, 392-393, fig. 262; Bartoloni *et al.* 1987, 142; Maaskant-Kleibrink, M. 1991, 86; Gallina & Malnati 1992, 23 (ill.); Naso 1996, 359 n. 557 and 401 n. 663.

13. The model is traditionally dated to the Sala Consilina period IB, i.e. the late 9th-early 8th century BC. However, P. Ruby places the tomb in the early phase IC and considers this phase parallel to Toms' Veii phase IC or Close-Brooks' end of Veii phase IB/phase IIA. This would indicate a date around 800-760 BC, which seems more convincing for the model (Ruby 1990, vol. I, 211, 216 and annexe 5).

14. For a discussion of the inclination of thatched versus tiled roofs, see Damgaard Andersen 2001.

15. Except for the Copenhagen and Louvre models such trap doors are only known from the Etruscan house urn from Cerveteri, Tomba della Nave III, dated 650-600 BC. In this case the roof-shaped lid is removable (only the lid is preserved), cf. Buranelli 1985, 58-59, figs. 28-29.

16. E.g. Gentili 1987, 242-257, esp. 243-246, fig. 162; Gentili 1988, 84, fig. 2; Merlo 1991; von Eles 1996, 56, 65, figs. 42, 44, 51 (note that the detailed photograph showing one of the huts on fig. 42 is reversed); Sassatelli 1996, figs. 17-18, tav. 1.

17. Cf. Russo Tagliente 1992 for description of the buildings and site bibliographies: site 3 (Amendolara), site 8 (Buccino), site 9 (Cairano), site 10 (Cancellara), site 18 (Garaguso), site 36 (Pisticci), site 38 (Ripacandida), site 39 (Roccagloriosa), site 45 (Scalea) and site 47 (Serra di Vaglio). To these should be added the three (possibly more) houses from the late 6th century at Palinuro (Naumann 1958). The Archaic phase of the building in loc. S. Stefano, Buccino (Russo Tagliente 1992, site 8) and the late 6th century three-room building in contr. Sotto le Quote, Garaguso (site 18), have not been considered as no details of these houses have been published. We have chosen to define a house as a building on a stone foundation. Huts are normally of a rounded ground plan. They have been reported at Cancellara, Pisticci, Ripacandida, Roccagloriosa and Serra di Vaglio; but some of them are dubious: Tagliente seemed reluctant to identify an hourglass-formed structure at Pisticci without postholes as a hut (Tagliente & Lombardo 1985, esp. 288), whereas Russo Tagliente (1992, 257) accepts it without reservations.

18. For this reason the typology of Archaic buildings in Russo Tagliente 1992, 88-89 seems little convincing.

19. An enigmatic exception are the maximum 12 cremation tombs from Palinuro reported by Sestieri, some in *pozzi* with "Oinotrian" grave goods, others in a tile *cassa* with few vases of Ionian type. Sestieri believed these to be the graves of Hellenized Oinotrians and Greeks respectively. For a recent evaluation of the material and references see Fiammenghi 1985, 7-16; a reinterpretation of the "Cremation tombs" from Palinuro in Horsnæs 1999, vol. 2 pp. 106-111. For examples from the Lucanian phase, see Gualtieri 1982.

20. For a discussion of the connection between the house and the individual/deceased see e.g. Toms 2001.

21. We thank dott.ssa Nava for her information on the models presented in October 1997.

22. Exhibition catalogue p. 218. The text provides no information of the context of the models.

23. *Archeo* 4, April 1999, 21-22, colour illustration on p. 22. We thank Dr. Annette Rathje for drawing our attention to this illustration.

24. We are most grateful for this personal communication from dott. Salvatore Bianco, who – together with dott.ssa Nava – is preparing a publication of all the models.

BIBLIOGRAPHY

Adamesteanu, D. 1970-1971
Una tomba arcaica da Armento, *AMSMG* 10-11, 83-92.

Adamesteanu, D. 1974
La Basilicata antica, Cava dei Tirreni.

Bartoloni, G., Beijer, A., De Santis, A. 1985
Huts in central Tyrrhenian area of Italy during the Protohistoric Age. BAR 245.

Bartolini, G. et al. 1987
Le urne a capanna rinvenute in Italia (Archeologica 68), Rome.

Bianco, S. et al. (eds.) 1996
I Greci in Occidente. Greci, Enotri e Lucani nella Basilicata meridionale. Napoli.

Bottini, A. 1981
Ruvo del Monte (Potenza) - Necropoli in contrada S. Antonio: scavi 1977, *NSc* 1981 (1982), 183-288.

Bottini, A. & Barra Bagnasco, M. 1990
L'attività archeologica in Basilicata 1989, *Atti del 29 Convegno di Studi sulla Magna Grecia*, Taranto, 551-577.

Brijder, H.A.G., Beelen, J. & van der Meer, L.B. 1989
De Etrusken, Exhibition Allard Pierson Museum, Amsterdam, October 1989 - February 1990. 's-Gravenhage.

Brunn, H. 1853
Notizie intorno alla collezione di antichità de' sigg. Amati a Potenza e Fittipaldi ad Anzi di Basilicata, *BdI*, 159-168.

BTCGI 1977-1996
Bibliografia Topografica della Colonizzazione Greca in Italia e nelle Isole Tirreniche I-XIV. Pisa - Roma.

Buranelli, F. 1985
L'Urna "Calabresi" di Cerveteri (Studia Archeologica 41). Roma.

Coen, A. 1991
Complessi tombali di Cerveteri con urne cinerarie tardo orientalizzanti. Biblioteca di "Studi Etruschi" 21. Firenze.

Colonna, G. 1986
Urbanistica e architettura, in: G. Pugliese Carratelli (ed.), *Rasenna. Storia e civiltà degli etruschi.* Milano, 392-393.

d'Agostino, B. 1974
La civiltà del ferro nell'Italia meridionale e nella Sicilia, *PCIA* 2, Rome.

Damgaard Andersen, H. 2001
Thatched or tiled roofs from the Early Iron Age to the Archaic period in Central Italy, in: L. Karlsson & R. Brandt (eds.), *Proceedings of the conference "From huts to houses - transformation of ancient societies"*, Rome, 22-24 September 1997, OpRom.

Danner, P. 1992
Tonmodelle von Naiskoi aus Kalabrien, *Rivista di Archeologia* 16, 36-48.

De Caro, S. & Borriello, M. 1996
La Magna Grecia nelle collezioni del Museo Archeologico di Napoli. Catalogo della mostra. Napoli.

Drerup, H. 1969
Griechische Baukunst in geometrischer Zeit (ArchHom O). Göttingen.

Edlund Gantz, I. 1972
The seated statue akroteria from Poggio Civitate (Murlo), *DialArch* n.s. 6,2/3, 167-236.

von Eles, P. 1996
Museo Civico Archeologico. Verucchio. Verucchio.

Fiammenghi, A. 1985
La necropoli di Palinuro: ricostruzione di una comunità indigena del VI sec. a.C., *DdA*, 3. ser. 3:2, 7-16.

Gallina M.A. & Malnati, L. 1992
Gli Etruschi e il loro ambiente (Popoli dell'Italia antica). Milano.

Gentili, G.V. 1987
Verucchio, in: Bermond Montanari, G. (ed.), *La formazione della città in Emilia Romagna.* Exhibition catalogue Bologna 1987-1988. Bologna, 207-263.

Gentili, G.V. 1988
Testimonianze dell'abitato villanoviano ed "etruscoide" di Verucchio, in: *La formazione della città preromana in Emilia Romagna*, Atti del Convegno di studi Bologna-Marzabotto 7-8 dicembre 1985. Imola.

Greco, G. 1980
Le fasi cronologiche dell'abitato Serra di Vaglio, in: *Attività archeologica in Basilicata, Studi in onore di Dinu Adamesteanu*, Mantova, 367-404.

Gualtieri, M. 1982
Cremation among the Lucanians, *AJA* 86, 475-481.

Hano, *et al.* 1971
Garaguso (Matera). - Relazione preliminare sugli scavi del 1970, *NSc*, 424-438.

Horsnæs, H.W. 1999
The Cultural Development in North-Western Lucania c. 600-273 BC. Ph.d.-dissertation, Copenhagen University.

Kilian, K. 1962a
Mostra della preistoria e della protostoria del Salernitano. Salerno.

Kilian, K. 1962b
Beiträge zur Chronologie der Nekropole Sala Consilina, *Apollo* 2, 81-103.

Kilian, K. 1970
Archäologische Forschungen in Lukanien 3 (RMErgH 15). Heidelberg.

La Genière, J. de 1968
Recherches sur l'Âge du Fer en Italie Méridionale, Sala Consilina, Naples.

Maaskant-Kleibrink, M. 1991
Early Latin settlement-plans at Borgo Le Ferriere (*Satricum*), *BABesch* 66, .

Malnati, L. 1984
Tombe arcaiche di S. Maria D'Anglona (scavi 1972-1973), *Quaderni di Acme 4*, 41-95.

Merlo, R. 1991
La capanna e il trono, *Archeo* 72, 43-49.

Morel, J.-P. 1970
Cronique, *MEFRA* 82, 556-652.

Nafissi, M. 1985
Le genti indigene: Enotri, Coni, Siculi e Morgeti, Ausoni, Iapigi, Sanniti, in: G. Pugliese Carratelli (ed.): *Magna Grecia. Il Mediterraneo, le metropoli, e la fondazione delle colonie*, Milano, 189-208.

Naso, A. 1996
Architetture dipinte. Decorazioni parietali non figurate nelle tombe a camera dell'Etruria meridionale (VII-V sec. a.C.) (Bibliotheca archaeologica 18). Roma.

Naumann, R. 1958
Palinuro, Ergebnisse der Ausgrabungen, I. Topographie und Architektur (RMErgH 3). Heidelberg.

Naumann, R. & Neutsch, B. 1960
Palinuro. Ergebnisse der Ausgrabungen II. Nekropole, Terrassenzone und Einzelfunde (RMErgH 4). Heidelberg.

Orlandini, P. 1980
Figura umana e motivi antropomorfi sulla ceramica enotria, *Studi in onore di Ferrante Rittatore Vonwiller II*, Como, 309-316

Pinza, G. 1905
Monumenti primitivi di Roma e del Lazio antico (MonAnt 15). Roma.

Pottier, E. 1897
Vases antiques du Louvre. Paris.

Pottier, E. 1899
Musée du Louvre - Catalogue des vases antique de terre cuite 2. Paris.

Richter, G.M.A. 1966
The furniture of the Greeks, Etruscans and Romans. London.

Ruby, P. 1990
Le crepuscule des marges. Etude chronologique et sociale de la nécropole du Premier Age du Fer de Sala Consilina - Italie. Diss., Paris. (published as: *Le crepuscule des marges: la premier âge du fer à Sala Consilina*. Bibliothéque des l'ecoles françaises d'Athenes e de Rome 290, Collection du Centre Jean Bérard 12. Naples 1995).

Russo Tagliente, A. 1992
Edilizia domestica in Apulia e Lucania, ellenizzazione e società nella tipologia abitativa indigena tra VIII e III secolo a.C. (Deputazione di Storia Patria per la Lucania, Quaderni di Archeologia e Storia Antica 4). Galatina.

Russo Tagliente, A. 1992-1993
Chiaromonte (Potenza) - La necropoli arcaica in località Sotto La Croce, scavi 1973, *NSc* 1992-1993 (1996), 233-407.

Sassatelli, G. 1996
Verucchio, centro etrusco "di frontiera", *Ocnus* 4, 261-265.

Schattner, T.G. 1990
Zur Entstehung des Dreieckgiebels, in: *Akten des XIII: internationalen Kongresses für klassische Archäologie* (1988), Berlin, 405-407.

Sestieri Bertarelli, M. 1958
Il tempietto e la stipe votiva di Garaguso, *AMSMG*, n.s. 2, 67-78.

Staccioli, R. 1968
Modelli di edifici etrusco-italici. I modelli votivi. Firenze.

Stopponi, S. (ed.) 1985
Case e palazzi d'Etruria (mostra Siena). Milano.

Strong, D. 1968
The Early Etruscans. London.

Tagliente, M. & Lombardo, M. 1985
Nuovi documenti su Pisticci in età arcaica, *PP* 223, 284-307.

Tocco Sciarelli, G. 1980
Aspetti culturali della Val d'Agri dal VII al VI sec. a.C., in: *Attività Archeologica in Basilicata, Studi in onore Dinu Adamesteanu*, Mantova, 439-476.

Toms, J. 2001
The hut and the house in Early Iron Age and Orientalizing Etruria, in L. Karlsson & R. Brandt (eds.), *Proceedings of the conference "From huts to houses - transformation of ancient societies", Rome, 22-24 September 1997, OpRom.*

Yntema, D. 1984
The Matt-Painted Pottery of Southern Italy. Amsterdam (Later edition with corrections and alterations, especially on the Bradano section. Lecce 1990)

Helle Damgaard Andersen
Tjørnevænget 20
DK-2800 Lyngby

hda.its@cbs.dk

Helle W. Horsnæs
Nordre Fasanvej 201, 5tv
DK-2000 Frederiksberg

kollhors@worldonline

ARCHAIC KARIAN POTTERY – INVESTIGATING CULTURE?

ANNE MARIE CARSTENS

"Caria to the south was not Greek, but its pottery – at least in the nearer parts – followed East Greek fashions." This is how R. M. Cook briefly described Karia and its Archaic pottery production (Cook 1998, 6). The statement remains unexplained, but to me it raises the questions of how the "Greekness" is recognised and, not least, how it is measured. How does the archaeological discipline measure a cultural group, if not by the artefacts produced by that group?

By tradition archaeological material is divided into typological groups defined either by geographical distribution or by chronological position. These groups are often labelled by the names of the ancient landscapes where they occur most frequently, or by the name of certain groups of people that we think may have lived there. Such divisions are created by the archaeologist, and we do not always know if such a division existed in antiquity. These are the normal premises for the definition of archaeological cultures: first and foremost, they are based on archaeological typologies and may not represent groups of real acting people.[1] We cannot hastily conclude that a vessel with an East Greek decoration was made or used by a Greek, and we can only offer hypotheses of whether the Greekness of the vessel Hellenized the owner. Speculations of this kind remain hypothetical, yet they will constitute the cornerstone of archaeology if we wish to continue the transformation of the study of ancient pottery from a study of artefacts *per se* into a study of the peoples who produced and made use of these artefacts.

In the following article I wish to point out the difficulties (or even pitfalls) that archaeological interpretation conducted on the basis of typological studies involves. It can be characterised as the latent conflict between material culture and cultural groups constituted by and consisting of people.

The emergence of Karian Archaic pottery

In reviewing the studies of Karian pottery from the Geometric and Archaic period – since the early findings in the 19th century until the latest attempts to define the Karian style as a specific sub-group of the so-called East Greek pottery – we may find an explanation as to the nature of Karian pottery. But, what is more, we may reach a more diversified characterisation of Karian culture. (**Fig. 1**)

In 1886, W. R. Paton excavated a number of Protogeometric and Geometric chamber tombs belonging to the ancient settlement at Assarlık in the southern part of the Halikarnassos peninsula. While the sepulchral architecture of the Assarlık tombs was always considered more "indigenous" than Greek, the pottery and other artefacts found in the tombs were immediately identified as being Greek imports, probably Attic, or as pieces produced by Attic immigrants (Paton 1887; Desborough 1952, 218-222). The settlement was, in other words, explained within the framework of migrations in the Early Iron Age. In the early 1960s, following the discovery of the Late BronzeAge necropolis at Müsgebi, the small Submycenaean cemetery at Çömlekçi between Bodrum and Milas, and the Protogeometric chamber tomb at Dirmil, the nature of the earliest Greek settlements and the Hellenization of south-western Karia, in particular, had to be reconsidered (Bass 1963; Boysal 1967). In 1971 A. Snodgrass concluded that the Assarlık pottery was probably made locally, although the earliest finds may have been imports (Snodgrass 1971, 67). The increasing amount of archaeological evidence from the Early Iron Age in south-western Anatolia illustrates that the Bronze Age breakdown was not as profound nor long-lasting as has so far been believed. Likewise the migration in the Early Iron Age had a much less significant effect and was of a more diffuse nature. The idea of continued habitation has in recent years gained a foothold, exemplified in the study of the Late BronzeAge and Protogeometric Miletos (Niemeier & Niemeier 1997).

C. Özgünel, who had participated as a student on the Müsgebi excavations, wrote his dissertation on the Karian Geometric pottery (Özgünel 1979). He worked mainly with finds from rescue excavations conducted near Milas, but also included Geometric material from the excavations in the vicinity of the chamber tomb at Dirmil. Around the time when Özgünel's monograph was published, R. M. Cook, who had always enjoyed the East Greek pottery styles, had begun working on

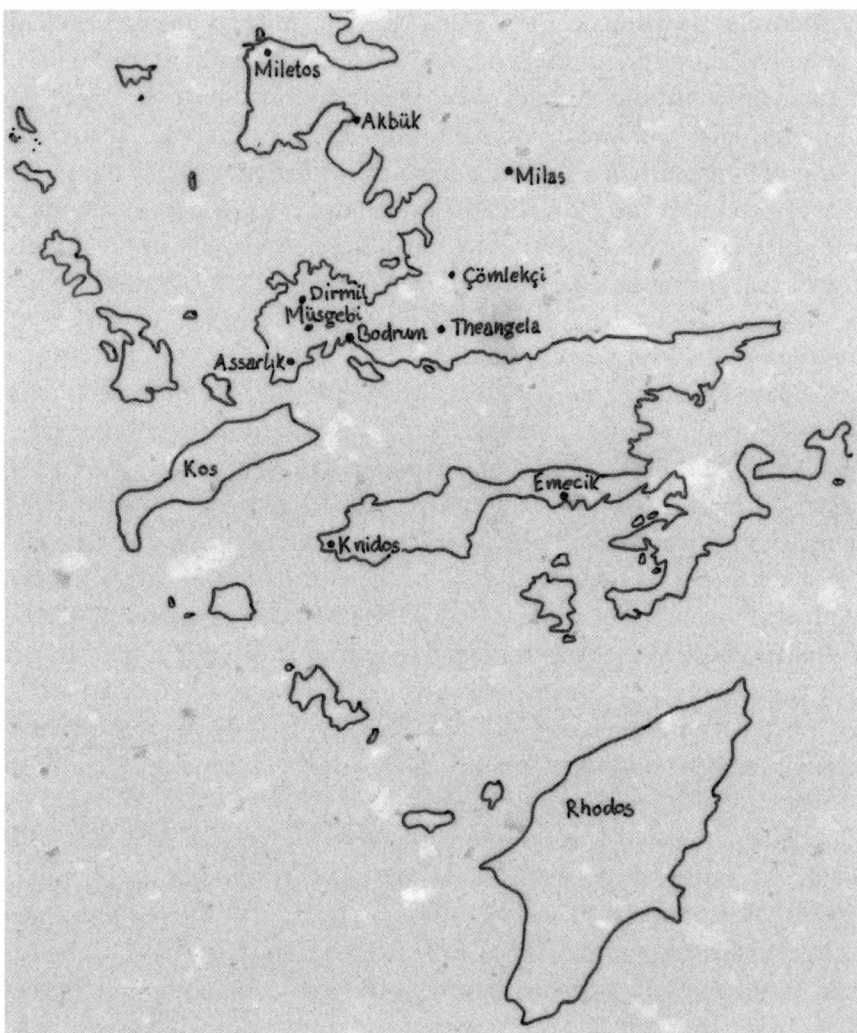

Fig. 1: Map of south-western Karia and the Dodekanese, showing the sites mentioned. The exact location of the Damlıboğaz necropolis is unknown to the author.

the later Karian Archaic pottery (Cook 1980). In 1980 a monograph appeared about a local production of terracotta figurines possibly located in ancient Theangela, east of Bodrum (Isik 1980), and later locally produced Geometric pottery was also found at the site (Isik 1990). Thus by the beginning of the 1980's the archaeological evidence had started to form a picture of southwestern Karia as a region producing pottery in the "Greek" style.

Although the presence of "Greek" pottery in local production may not indicate any special Hellenization of the people, it indicates a level of contact with the Aegean world, which must have been strong enough to make an impact on the local craftsmen. It speaks of a wider world of contact than a few concentrated settlements would have been able to produce. The "Greekness" must have been widely interwoven into the style of the Western Anatolian artefacts already in the mature Geometric period (see e.g. Osborne 1996, chapter 2; Özgünel 1979).

Excavating Archaic Karian pottery
In the late 1960's a large cemetery at Damlıboğaz, between Milas and Miletos was plundered. The finds were later distributed among a large number of Middle European and American museums, and were published during the 1980's – mainly by R. M. Cook. A number of other scholars, in particular D. Lenz, have contributed to the study of the Karian Archaic pottery and, not least, to the elucidation of their undocumented archaeological context (Cook 1980, 1993, 1999; Lenz 1997).
The Damlıboğaz case is unfortunately not an isolated incident. In the past twenty years Archaic tombs at Theangela, east of Halikarnassos, have also been plundered (Isik 1990), while only sporadic legal excavations have been conducted, mainly in the form of rescue excavations in connection with building activities.

Scientific fieldwork concerning the Archaic period is, at the moment, concentrated especially on Miletos and its hinterland (Niemeier 1997, 1999; Lohmann 1997, 1999). The excavations at the site presumed to be the Triopion sanctuary at Emecik (Old Knidos) may, in the future, produce more Archaic Karian pottery in good contexts. (Berges 1996; Berges & Tuna 2000). From these excavations it will hopefully be possible to distinguish not only the relative chronology but also the distribution of production centres of Archaic Karian pottery. The largest existing group of Karian pottery still consists of vessels bought on the art market, presumably deriving from the looted necropolis at Damlıboğaz (although the commercial success of this operation may have resulted in other clandestine activities). These vessels form the core of the material that has enabled Cook to establish a rather well defined sub-group of the East Greek pottery, Karian Archaic pottery.

Karian pottery from Geometric to Archaic

Karian Geometric pottery did indeed follow the Greek fashions, but in its own Karian way. Though the Protogeometric material from southwestern Karia is meagre, Özgünel described the Late Protogeometric style as rather heavily influenced by Attic pottery – but with clear local characteristics. While the repertoire of shapes match the Attic repertoire, the decoration and its arrangements may represent a local style. A number of finds from the Dirmil chamber tomb illustrate this.

On a skyphos a simple careless zigzag line connects the concentric semicircles (Özgünel 1979, cat.no. 6).[2] A belly-handled amphora is, in the main shoulder panel, decorated with a series of circles with a double chain of lozenges drawn between them (Özgünel 1979, cat.no. 2).[3] A very impressive krater has often been connected to the Marmariana kraters of Thessaly. While both shape and decoration display rather close similarities, the piece is by and large an exception from the otherwise massive Attic influence (Özgünel 1979, 69).

For Özgünel there was little doubt that the Karian pottery style was fostered, if not born, from the early Protogeometric period. Already in the Early Iron Age the settlements in Karia were self-sufficient with regard to pottery and they created their own style as a result of, and as a response to, inspiration from other regions. The Dodekanese influence, which was quite extensive in the Late Bronze-Age material from the Müsgebi necropolis (Carstens 1999, 24-28), now declined. Mirroring Karia's geographical position, the Early and Middle Geometric pottery was composed of both Attic, Argive, and Dodekanese elements. A neck-handled amphora with hatched zigzag decoration on the metope neck panel illustrates this combination of ingredients quite well (Özgünel 1979, cat.no. 8). While the shape agrees with the Attic amphorae e.g. from the Kerameikos, the decoration as a whole may find parallels in Argive pieces. The maeanders have Attic counterparts, and the hatched zigzag and the chain of hatched lozenges can be found in Dodekanese pottery (Özgünel 1979, 76-77).

The amount of Karian pottery increases in the Late and Sub-Geometric period and most of the material derives from rescue excavations near Milas (Akarca 1971). While the different sources of inspiration remains the most characteristic feature, the "Karianness" of the pottery, especially when it comes to decoration, derives from elements that seem more like interpretations than imitations – they represent conscious choices and demonstrate a clear local sense of innovation. A kantharos, found near the chamber tomb at Dirmil, has applied writhing snakes on the handles. The piece is unique in Western Anatolian pottery, even though the kantharos was presumably produced locally (Özgünel 1979, cat.nos. 32, 85, 95). We may be able to detect the different styles that had an impact on Karian Geometric pottery. However, there can be little doubt that Karian pottery production was firmly established by the Late Geometric period – and the stage was set for the richly decorated Archaic styles , that is the Karian Archaic pottery.

The Archaic pottery of Karia
The material evidence, which enabled R. M. Cook to define the Archaic Karian pottery as a sub-group of the East Greek pottery, consists largely of whole vessels purchased mainly by Middle European Museums and private collectors during the 1960s and 1970s (Hemelrijk 1987, 33). One of the largest and most representative collections is now in Kassel (Gercke 1981), while the Martin von Wagner Museum in Würzburg also has a fine collection of Karian pottery (Lenz 1997).[4]

The earliest group of Archaic Karian pottery from the middle of the seventh century BC consists mainly of broad oinochoai. The depressed shape is emphasised by the decoration, which mainly covers the shoulder. The lower part of the vase is only decorated with bands or a wide halo. Divided into small panels by vertical lines are chessboard patterns and hatched maeanders. The figures, goats or the very popular geese, usually occur alone, rarely in groups (e.g. Gercke 1981, cat.no. 5) (**Fig. 2**). The strong geometric, almost subgeometric style, the taste of the hatched patterns and the chessboard (often hatched as well) may reflect Phrygian pottery (Lenz 1997, 35). In general the ornaments are rough and rather big, the stroke broad and, at times, careless.

In the early sixth century BC the Wild Goat vases enter the scene. In this group of vessels, too, the depressed shape and the shoulder zone are emphasized. Rarely is the lower belly-zone decorated in any detail. An impressive dinos from Kassel shows the strong geometric influence, by, amongst other characteristics, the chequered patterns in the upper handle zone (Gercke 1981, cat.no. 33) (**Plate 10**). The central decoration of the handle zone is heraldic with a palmette as the centre on the one side, on the other a lotus. Out of the palmette grows a peculiar creature, a siren with a crocodile's mouth. It faces a bull marching towards

Fig. 2: Squat jug with panel decoration. 660/650 BC. Staatliche Kunstsammlung Kassel, inv.no. Alg. 267 (from Gercke 1981, 34).

the centre of the decoration. Between the siren and the bull, a goose is preening itself. Behind the bull is a goat and above the back of the bull a dog is running rapidly, also towards the centre. An almost exact copy of this scene is repeated on the other side of the central double palmette. On the other side of the vase, the central double lotus is placed in a frame together with eagles, which flank the lotus as they are flying towards it. Outside the frame, goats occupy the space between the frame and the handles. On this side of the vase too, the accuracy of the mirrored motives is remarkable. Because of this mirrored symmetry, the animals that flank the central decoration appear more like ornaments than beasts.

In the later Wild Goat group, Cook has identified a painter, the Bochum Painter (Cook 1993, 1999, 89). This Painter frequently used the broad bodied oinochoe with two-reed-handles. Here too, much emphasis is laid on the shoulder zone, whatever the shape of the vessel. This creates a heavy impression that is underlined by the decoration. The panels on the vessel are framed by cables or maeanders, the belly often has a "keyboard" ornament while the lower part is only decorated with bands. In general the ornaments are large and somewhat sloppy. The animals, goats and geese, are roughly drawn, but with a certain vitality and, certainly, not without charm (Bochum S 988) (**Fig. 3**).

While the Karian Wild Goat variations are inspired by and related to the south Ionian production the local tradition is strong. A wide variety of filling ornaments, most typically the keyboard ornament and bands of sickles on the belly, characterises the decoration though the style is somewhat rough and at times sketchy. The repertoire of vessel shapes, where the oinochoai seem quite dominant, may reflect the funerary context rather than the range of shapes produced.

Fikellura pottery is also included in the Archaic Karian production, both the richly ornamented ramification of the Wild Goat pottery, and the free field Fikellura, best known from the Running Man painters. A jug at Kassel may serve as an example of the first group(Gercke 1981, cat.no. 4). This squat jug has a sloppy halo surrounding the broad base, and the body is decorated by bands and hourglass ornaments. Placed on the shoulder, the figural decoration is divided into three main panels by chequered triangles. The central and biggest panel is occupied by a goat framed by vertical lines. On each side, outside this secondary frame, a goose is turned towards the goat. The two side panels also contain a goose, but this time turned towards the band handle. The neck is decorated with another chequered panel.

The free field Karian Fikellura is found both in a variation where the figures are drawn in contour and also in the almost solid black style of the Running Man vases (Gercke 1981, cat.no.3). One rather peculiar example of the first type is an amphora in Bern (Cain 1974). The shape is the familiar low Fikellura amphora, typical of this style. Only the upper part of the shoulder is decorated with the classic chain of buds. Just above the base, two bands serve as a frame. One figure, drawn in outline, covers each side of the amphora. On the one side a bull, on the other a lion are drawn in a careless yet elegant line. Their limbs are long and

Fig. 3: Oinochoe, Antikensammlung der Ruhr-Universität Bochum, inv.no. S988. Photo by courtesy of the Antikensammlung.

slender and the lion especially has an extremely curved spinal line. Irregular V-shaped lines are used for marking the coat of the animals.

This type of outline drawing is also seen in the earliest group of Karian Wild Goat pottery, for example, in the form of small animal protomes seen both on a jug in Utrecht (Hemelrijk 1987, cat.no. 1) and on a jug in Kassel (Gercke 1981, cat.no. 6) (**Fig. 4**). In both cases the protome may best be interpreted as representing a lion.

In *Greek Painted Pottery* Cook described the Fikellura pottery in a few words, as possessing "a shrewd originality" (Cook 1997, 124). The originality of the Archaic Karian pottery does not seem, at least in the

Fig. 4: Squat jug with lion in outline drawing. 660/650 BC. Staatliche Kunstsammlung Kassel, inv.no. Alg. 247 (from Gercke 1981, 35).

developed Fikellura period during the 560s and 550s, less impressive. The animals are often extremely vivid and the carelessness of the strokes enchanting. Although the local Karian pottery production reflected part of a general trend, it expressed, with some emphasis, its independence: "Refreshing, unpremeditated reinterpretation of Orientalizing formulas" (Cook 1999, 91). Variations within a known scheme – this is how Cook has tried to catch some of the "Karianness" of the pottery. Let us, with this description of the Karian pottery styles in mind, return to the Karian culture.

Greek and Karian

E. A. Hemelrijks publication of seven Karian vases from Dutch collections (Utrecht and Amsterdam) included a discussion on the acculturation between Greeks and Karians. Indeed, it seems an important occupation for archaeologists working in Karia to find and try to establish a satisfactory description and definition of the Karians.[5] The Karians are often understood as being a group of (local) people set apart from Greeks: "Culturally the Greeks were superior and Carians living in the Greek cities on the coast seem to have adapted themselves to Greek culture thoroughly." (Hemelrijk 1987, 54).

H. Lohmann stated recently while comparing the rural buildings in the hills north of Akbük on the Milesian peninsula with the same type of buildings on the Bodrum peninsula: "Hier wie dort handelt es sich um die ständigen Wohnsitze von Hirtensippen und hier wie dort stand die karische Hirtenbevölkerung offenbar in einem engen wirtschaftlichen Austausch mit der griechischen Bevölkerung, denn die Keramik in den Gräbern und Compounds ist in beiden Regionen rein griechisch."[6]

Literary sources mention both Karians and Greeks living in ancient Karia.[7] In this respect the distinction is an ancient one, often understood as if two separate ethnic groups were living there (Hornblower 1982, Part I; Carstens & Flensted-Jensen forthcoming). An ethnic group should, in the traditional archaeological concept of ethnic differentiation, be characterised as a different archaeological culture, consisting of a distinct and characteristic group of objects (Jones 1997, 15-39). However, both quotations above illustrate the difficulties archaeology is facing when trying to match archaeological cultures with peoples. We find pottery of the archaeological culture X inside dwellings or tombs belonging to the culture Y. But we forget that these material cultures may not be any more qualified to receive the label of cultures than, say, yoghurt. They present groups of objects organised by the archaeologists, labelled X or Y, Greek or Karian. The Karian culture is then the heading for a group of objects. But it is applied as if it included people organised in a similar way. An archaeological culture is a well defined selection of objects that are regarded as being typical of a given area or a given period, or both. Such a concept of culture is an operational concept and in its extreme form only explains its own construction. Archaeological analyses and interpretations that spring from such a position easily become one-dimensional and categorical.

C. Geertz has once, quoting Weber, characterised culture as the web, which entangles man (Geertz 1973, 5). I prefer to explain the situation in this way: if culture is the web, then the objects that archaeology studies are traces of the web – like the dots in a child's painting book, only here they are not numbered. We have to add the lines and figure out their interconnection. In my opinion, a "human" concept of culture must remain holistic and indeterminate. It is impossible (and meaningless) to define, since it consists of a notion of inter-human relations, which archaeology may only find represented in the material record of these encounters or not. Thus, instead of searching for the definition of an ultimate concept of culture which then should be both operational (like a check list) and holistic (all-encompassing explanation), it would perhaps be more fruitful to describe and characterise the archaeological evidence in its archaeological, topographical and historical context. In other words, in its cultural context, even if we are not certain what this notion involves (Møllgaard 1992).

This ,generally speaking, is what Classical Archaeology is occupied with. However, the traditional trend in the study of pottery is only of minor interest in contextual studies: the Karian Archaic pottery is described and defined "on its own", on the basis of a number of vessels and regardless of their archaeological context.[8] The school of Karian pottery – vessels sharing stylistic features – is used as an argument in the reconstruction of cultural interrelations, at times overtly interpreted: East Greek and Karian pottery are seen as distinct cultural manifestations reflecting similarly distinct human behaviour. Thus the Greek pottery inside Karian dwellings is explained by a supposed interrelation brought about by the Karians supplying the Greeks with wool in exchange for pottery.[9] I find it more convincing that the mix of Karian and Greek goods found there is a result of coexistence, a sign of a close interrelation, of a successful acculturation. Rather than imagining two distant cultures (and people), operating on each side of a sharply drawn cultural borderline, I see the differences as evidence of a shaded border zone between a group of shepherds and a group of city dwellers. The Greekness of the Karians and the Karianness of the Greeks in Karia may even be what we see reflected in the Karian Archaic pottery.

Objects in this group are labelled Karian since they were found and, probably, produced there. There was a relation of both dependence and independence with regard to the other East Greek ceramic "schools", in particular the south Ionian.[10] We are left without an answer to the question: who produced the Karian pottery, Greeks or Karians? I would suggest that it does not matter. It was probably produced in Karia, maybe at Milas (Hemelrijk 1987; Cook 1993), and maybe also somewhere else (Cook 1998, 66), and it represents one element of Karian culture. This Karian culture cannot be defined as a construction consisting of some specific elements. Rather it is a blend of something Greek, and something local, something Phrygian, something Dodekanese: a blend, which ends up as Karian. The style reflects the interrelations which existed between these regions in the Archaic period. A culture is a product of its interrelations; these are what made Karia Karian.

NOTES

1. The ethnic group is an "acting entity" or a group of "acting people". It is constituted by self-identification and this group formation is best described as "categories of ascription and identification by the actors themselves" (Barth 1969, 10). Inspired by social anthropology, a way of rethinking the traditional archaeological categories has been suggested by the Anglo-American school of theoretical archaeology, e.g. Shennan 1989. The issue is not whether ethnic groups can be identified in the archaeological record as such, but that the ethnic groups and their manifold formations make up a much more complex system of relations than archaeology ever included in its formation of ideas. A clear distinction between archaeological culture and real acting entities must always be stressed. We cannot expect ethnic groups to create manifestations, which can be detected in the archaeological material. The internal organisation of a given area that may be based on such ethnic differentiation will often remain unknown in the archaeological research. For a fuller discussion of these aspects: Jones 1997; Carstens 1999, 179-182 et passim; Carstens & Flensted-Jensen forthcoming.

2. "This placing of the motif is local composition for Western Anatolia, and especially for Caria." Özgünel 1979, 72.

3. Sherds from skyphoi with similar decoration were found during the Danish excavations at Vigli on Rhodes. Sørensen 1992, 29, 35.

4. Cook 1999 presents a list of the hitherto known Karian vases in Turkish, European and American collections. See also Hemelrijk 1987; Evren 1991; Evren 2000.

5. E.g. Bean & Cook 1955; Radt 1970; Hornblower 1982. See also Carstens 1999, Chapter 5; Carstens & Flensted-Jensen forthcoming.

6. Lohmann 1999, 465. It is not clear whether the pottery found in these buildings belongs to the Archaic Karian pottery described in this article or whether the term "rein griechisch" refers to pottery produced in Ionia, for instance Miletos. However, pottery from the Archaic tumulus tombs of the Halikarnassos peninsula belong to the Karian Geometric and Early Archaic pottery production, as defined by Özgünel and Cook. Carstens 1999, Chapter 2.2.

7. For example Thukydides I.iv; Herodotos I.171; Strabo 7.7.2; 13.1.58-59; 14.2.27.

8. Cook hardly considers the question of contexts – only as a reflection on the chronology the lack of "useful" contexts is remarked. Cook 1999, 92.

9. The dwellings Lohmann refers to are probably shepherd's pens. Radt 1992.

10. The same relations between Karia and South Ionia are also found in monumental architecture of the Classical period. Pedersen 1994.

BIBLIOGRAPHY

Akarca, A. 1971
Beçin, *Belleten* 35, 3-37.

Barth, F. 1969
Introduction, in: Barth, F. (ed.); *Ethnic groups and boundaries, the social organization of culture difference.* Oslo, 9-38.

Bass, G.F. 1963
Mycenean and Protogeometric Tombs in the Halicarnassus Peninsula, *AJA* 67, 353-361.

Bean, G.E. & Cook, J.M. 1955
The Halicarnassus Peninsula, *BSA* 50, 85-171.

Berges, D. 1995-1996
Knidos und der Bundesheiligtum der dorischen Hexapolis, *Nürnberger Blätter zur Archäologie* 12, 103-120.

Berges, D. & Tuna, N. 2000
Das Apollonheiligtum von Emecik, *IstMit* 50 171-214.

Boysal, Y. 1967
New Excavations in Caria, *Anadolu* 11, 31-56.

Boysal, Y. 1985
Einige neue Gefässe der Dark Ages aus Karia, in Kandler, M. et al. (eds.), *Lebendiges Altertumswissenschaft. Festgabe zur Vollendung des 70. Lebenjahres von Hermann Vetters.* Wien, 16-18.

Boysal, Y. 1989
Eine vorarchaische Oinochoe im Museum von Fethiye, in: Cain, H-U, Gabelmann, H. & Salzmann, D. (eds.), *Festschrift für Nikolaus Himmelmann.* Mainz am Rhein, 79-82.

Cain, H.-U. 1973-74
Eine neuartige, kleinasiatische "Fikellura-Vase", *JkBernHistMus* 53-54, 43-56.

Carstens, A.M. 1999
Death Matters. Funerary architecture on the Halikarnassos peninsula. Unpublished Ph.D.- Dissertation, University of Copenhagen.

Carstens, A. M. & Flensted-Jensen, P. forthcoming
Halikarnassos and the Lelegians, in: Isager, S. & Pedersen, P. (eds.), *The Salmakis Fountain and Hellenistic Halikarnassos (Halicarnassian Studies IV).* Odense.

Cook, R. M. 1978/80
Antecedents of Fikellura, *Anadolu* 21, 71-74 (Festschrift Akurgal).

Cook, R. M. 1993
A Carian Wild Goat Workshop, *OJA* 12, 109-115.

Cook, R. M. 1997
Greek Painted Pottery. London.

Cook, R. M. 1998
East Greek Pottery. London.

Cook, R. M. 1999
A List of Carian Orientalizing Pottery, *OJA* 18, 79-93.

Desborough, V.R.d'A. 1952
Protogeometric Pottery. Oxford.

Evren, A. 1991
Tire Müzesi'ndeki Karia Kökenli Kaplar, *TürkAD* 29, 193-224.

Evren, A. 2000
Efes Müzesi'nde Ki Karya Bölgesi Kapları. Efes 2000.

Geertz, C. 1973
The Interpretation of Cultures. London.

Gercke, P. 1981
Funde aus der Antike. Sammlung Paul Dierichs Kassel. Kassel.

Hemelrijk, E. A. 1987
A Group of Provincial East-Greek Vases from South Western Asia Minor, *BAbesch* 62, 33-55.

Hornblower, S. 1982
Mausolus. London.

Isik, F. 1980
Die Koroplastik von Theangela in Karien und ihre Beziehungen zu Ostionien zwischen 560 und 270 v. Chr. Tübingen.

Isik, F. 1990
Frühe Funde aus Theangela und die Gründung der Stadt, *IstMit* 40, 17-36.

Jones, S. 1997
The archaeology of ethnicity. London.

Lemos, A. A. 1991
Archaic pottery of Chios: the decorated styles 1-2. Oxford.

Lenz, D. 1997
Karische Keramik im Martin von Wagner-Museum, Würzburg, *ÖJh* 66, 29-61.

Lohmann, H. 1995
Milet 1992-1993. Survey in der Chora von Milet, *AA*, 293-329.

Lohmann, H. 1997
Survey in der Chora von Milet. Vorbericht über die Kampagnen der Jahre 1994 und 1995, *AA*, 285-311.

Lohmann, H. 1999
Survey in der Chora von Milet. Vorbericht über die Kampagnen der Jahre 1996 und 1997, *AA*, 349-473.

Møllgaard, J. 1992
Om kulturbegrebet, *Nordnytt* 45, 52-58.

Niemeier, B. & Niemeier, W-D. 1997
Milet 1994-1995. Projekt "Minoisch-Mykenisches bis Protogeometrisches Milet": Zielsetzung und Grabungen auf dem Stadionhügel und am Athenatempel, *AA*, 189-248.

Niemeier, W.D. 1999
"Die Zierde Ioniens" Ein archaischer Brunnen der jüngeren Athenatempel und Milet vor der Perserstörung, *AA*, 373-413.

Özgünel, C. 1976
Dirmil'de (Gökçebel) Bulunmuş Geometrik Kaplar, *Belleten* 40, 49-53.

Özgünel, C. 1979
Carian Geometric Pottery. Ankara.

Özgünel, C. 1977
Spätgeometrische Gefässe aus Karien, *AA*, 8-13.

Osborne, R. 1996
Greece in the Making. London.

Paton, W.R. 1887
Excavations in Caria, *JHS* 8, 64-82.

Pedersen, P. 1994
The Ionian Renaissance and some aspects of its origin within the field of architecture and planning, in: Isager, J. (ed), *Hekatomnid Caria and the Ionian Renaissance (Halicarnassian Studies I).* Odense.

Radt, W. 1970
Siedlungen und Bauten auf der Halbinsel von Halikarnassos unter besonderer Berücksichtigung der archaischen Epoche. Tübingen.

Radt, W. 1992
Lelegische Compounds und heutige verwandte Anlagen, in: Schütte, A. (ed), *Studien zum antiken Kleinasien* II (*Asia Minor Studien* 8). Bonn.

Shennan, S. 1989
Introduction: archaeological approaches to cultural identity, in: Shennan, S. (ed.), *Archaeological approaches to cultural identity.* London, 1-32.

Snodgrass, A. 1971
The Dark Age of Greece. Edinburgh.

Sørensen, L.W. 1992
Lindos IV.2. Excavations and Surveys in Southern Rhodes: Part 1. The Post-Mycenean Periods until Roman Times. Copenhagen.

Anne Marie Carstens
Gammel Køge Landevej 247
DK-2650 Hvidovre

amclfh@image.dk

Unless otherwise stated the illustrations are by the author

THE MEDITERRANEAN AND CENTRAL EUROPE IN THE 6TH AND 5TH CENTURIES BC. THE TRADE-ROUTE THROUGH THE RHÔNE VALLEY IN THE LIGHT OF DISCOVERIES OF LOCAL PLAIN WARES

LONE LEEGAARD

Finds of Attic pottery, as well as of bronze vessels of both Greek and Etruscan origin demonstrate that as early as the 8th century BC, the cultures of the Mediterranean were on trading terms with the developing cultures of Central Europe. This is particularly true of the period in question which is the 6th and 5th century BC (Pare 1991, 190-191, 197; Maffre 1995, 64-65). In this study we shall be focusing on areas of the *Hallstatt Culture* (700-475 BC) along the Rivers Saône, Doubs and Rhône north of Lyons and areas of the *La Tène* Culture (475-20 BC) along the Rhine (Kimmig 1983, fig. 46; **Fig. 1**).

The traditional view was that Greeks and Etruscans did not only produce these wares, but also retailed it in a sort of mutual competition on the Central European market, and that Greek merchants tended to use the River Rhône as their trade-route, while the Etruscans tended to favour a route across the Alps (Jully 1957; Joffroy 1960; Benoit 1965). The archaeological research of recent years has adapted itself to the general debate on the function and status attributed to e.g. Attic figured pottery in Antiquity. In consequence focus is not exclusively on the items mentioned above, but is also to an increasing degree concentrated on the less prestigious finds from areas along the trade-routes (Arcelin 1984 and 1986; Aigner-Foresti 1992; Pauli 1992; Trément 1999). These studies indicate that the exchange of goods was not merely a result of direct contact between the manufacturer and the consumer, but, to a greater extent, was due to a more diverse cultural development along the trade-routes in the areas discussed.

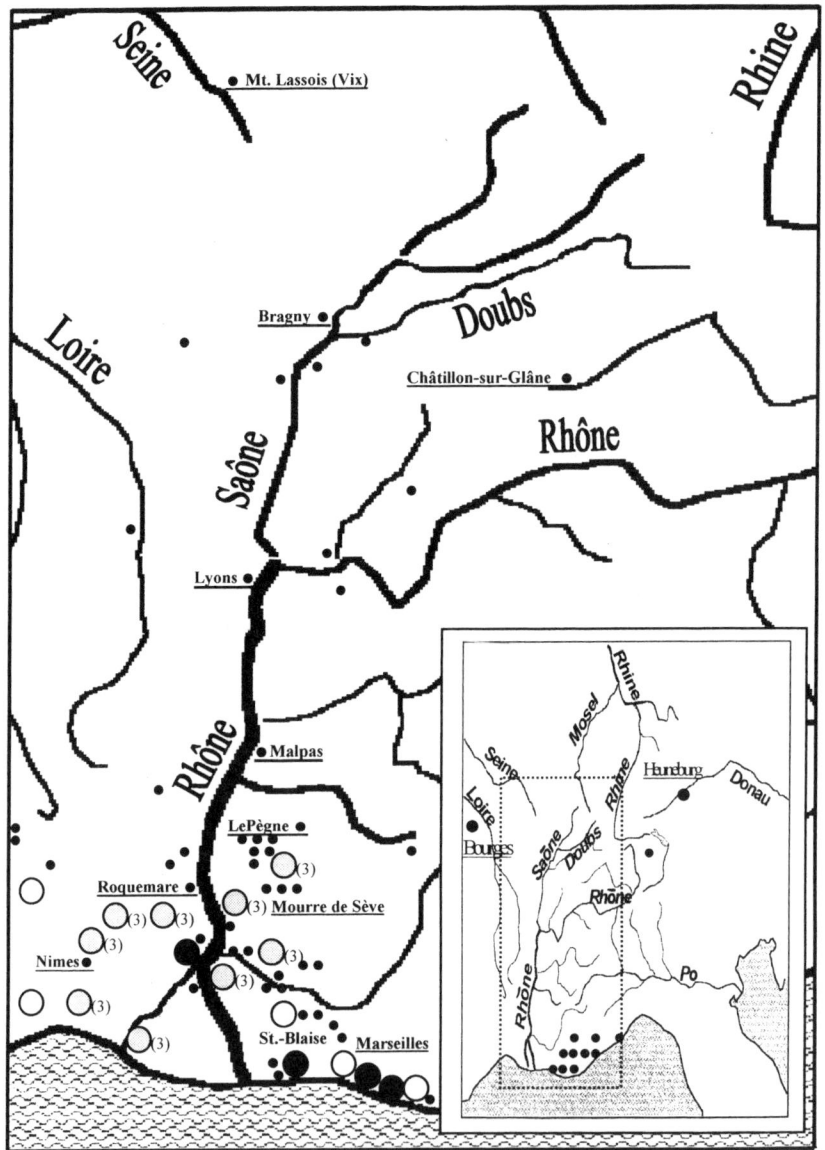

Fig. 1. Distribution of finds of pseudo-Phocaean pottery from the north coast of the Golfe du Lion, the Rhône Valley and Central Europe (Based on the author's observations).
More than 300 fragments found: Black circle. 100-300 fragments found: Grey circle. 30-100 fragments found: White circle. Less than 30 fragments found: Black dot. Only fragments classified according to shape are included (Figure 3). This excludes two thirds of the total amount of pseudo-Phocaean pottery from Languedoc (Gard). Find quantities, dominated by fragments from Arcelin's Group 3, are marked (3).

THE MEDITERRANEAN AND CENTRAL EUROPE

Sites	Pseudo-Phocaean pottery	Pseudo-Ionian pottery	Massaliot amphorae	Céramique cannelée	Attic figured. ware	Etruscan pottery	Etruscan and other Bronzeware
Ambérieau							
Asperg							
Bourges		(2)	(44)		(c. 48)		(10)
Bragny-sur-Saône	(> 3)				(> 3)	(a)	
Britzgyberg					(> 4)		
Breisach							
Camp-du-Chassey					(2)		
Charnes (Saône-et-Loire)							
Chateau-sur-Salins	(> 5)		(c. 20)		(> 11)		
Chatillon-sur-Glâne	(16)	(c. 4)	(c. 25)	(c. 100)	(c. 42)		
Heuneburg	(2)	(4)	(c. 15)		(c. 80)		
Jaevres (Loire)							
Lyons		(2)				(a)	
Malpas (Soyons)						(a)	
Mantoche			(c. 4)				
Mercey-sur-Saône							
Milly-Lamartine							
Mont Guérin							
Mont Lassois (Vix)	?		(c. 20)		(c. 100)	? (b)	
Montmorot	(2)						
Savoyeux / Seyssel (Ain)							
Tournus (Saône-et-Loire)							
Verjux							
Üetliberg	?						

Fig. 2. Overview of the juxtaposition of objects with Mediterranean associations, found in dwellings in Central Europe (Based on the author's observations).
(Lyons is the southernmost locality included). (Numbers given indicate the individual totals of finds, whether complete vessels or fragments).

SHAPES	PSEUDO-PHOCAEAN POTTERY Total amount studied (Group 3)	PSEUDO-IONIAN POTTERY Total amount studied (Type B)
A. Bowls	2270 (549)	908 (25)
B. Mediterranean types	914 (148)	3751 (289)
C. Biconical jars / jugs	181 (22)	72 (45)
D. Miscellaneous	7	149 (12)

Fig. 3. Overview of the distribution of the prototypes of the shapes in pseudo-Phocaean and pseudo-Ionian pottery (Based on the author's observations).
(The figures in parentheses indicate the number of fragments specifically characterized as pseudo-Phocaean Group 3 (Group 3), or pseudo-Ionian with subgeometric decoration (type B) respectively. The numbers indicate the proportion of the total number of finds given.)
A. Shapes common to Central-European and Mediterranean pottery (most frequently as Fig. 5: I-III, VI).
B. Shapes unique to Mediterranean countries (most frequently as Fig. 5: IV, V, VIII, X).
C. Shapes predominantly known within traditional Central-European pottery (most frequently as Fig. 5: VII & IX).
D. Other shapes.

Against this background, I shall discuss the structure of the trade-links through the Rhône Valley, basing my arguments on a wider range of archaeological evidence. Of prime significance in this context is the locally-produced plain wares from the Rhône Valley, i.e. the so-called pseudo-Phocaean and pseudo-Ionian wares. The discussion, however, also includes this kind of pottery from Marseilles, as well as Central European finds of this particular plainware, Attic pottery, Etruscan pottery, and bronzeware.

The distribution of these wares is given in **Figs. 2** and **3**. **Fig. 2** covers finds from Central Europe only, whereas **Fig. 3** covers finds from Provence and Central Europe. **Fig. 2** only registers Attic figured pottery when found in dwellings together with pseudo-Phocaean pottery, pseudo-Ionian pottery, Massaliote amphoras or the so-called *céramique cannelée*. Equally the bronze vessels are only registered when found in tombs clearly linked to dwellings with pseudo-Phocaean pottery, pseudo-Ionian pottery, Massaliote amphoras or *céramique cannelée*. The Etruscan pottery registered covers both finds of transport amphoras which are indicated with an **(a)** as well as Etruscan bucchero which is indicated with a **(b)** (Gran-Aymerich 1992, 340, note 52).

Fig. 3 registers the frequency of the most common shapes found among pseudo-Phocaean and pseudo-Ionian fragments respectively. Therefore the numbers given only cover fragments classified according to shape.

Pseudo-Phocaean Pottery
Pseudo-Phocaean pottery is an *unpainted* plain ware, produced in the Rhône Valley in the area south of Lyons in the period circa 550-400 BC (**Fig. 4**). This includes Marseilles, which was founded circa 600 BC by colonists from the East Greek city of Phocaea (Justin 43.4-12): hence the appellation "Phocaean" for the East Greek model, which is a grey bucchero appearing in a long series of variations as regards shape, decoration and treatment of the surface. It was imported into Marseilles, principally during the first half of the 6th century BC (Lamb 1932, 2-3, figs. 1-4; Arcelin 1984, 5-6, note 4, 142, notes 293-294).

The pseudo-Phocaean pottery from the Rhône Valley resembles its East Greek model as regards both the clay and method of manufacture. Thus in the majority of cases use was made of a potter's wheel and controlled firing, techniques which were unknown in the Rhône valley before the 6th century BC (Benoit 1965, 154). It also resembles its East Greek model from an aesthetic point of view, though without the latter's richness of variation in shape and decoration.

The shapes used are listed in **Figs. 5** and **3** which demonstrate that ten different shapes can be enumerated, most of which are "open," i.e. cups and bowls (**Fig. 5**: I-VI). Decoration is simple and is always based around the same motif: a circumferent band of incised lines which might be even or wavy. The surface of each vessel is polished, which derives from the fact that the original model was a pottery aspiring to a metallic appearance (**Fig. 4**).

Charlette Arcelin-Pradelle has undertaken a systematic analysis of all the available pseudo-Phocaean pottery discovered in the lands bordering the Rhône south of Lyons, i.e. the Rhône Valley. The findspots in the area can be divided into two regions as follows: the first is in Languedoc (Gard) to the west of the Rhône where Arcelin-Pradelle registers a total of 2012 pseudo-Phocaean fragments, of which less than one third can be classified according to shape (Arcelin, Dedet and Py

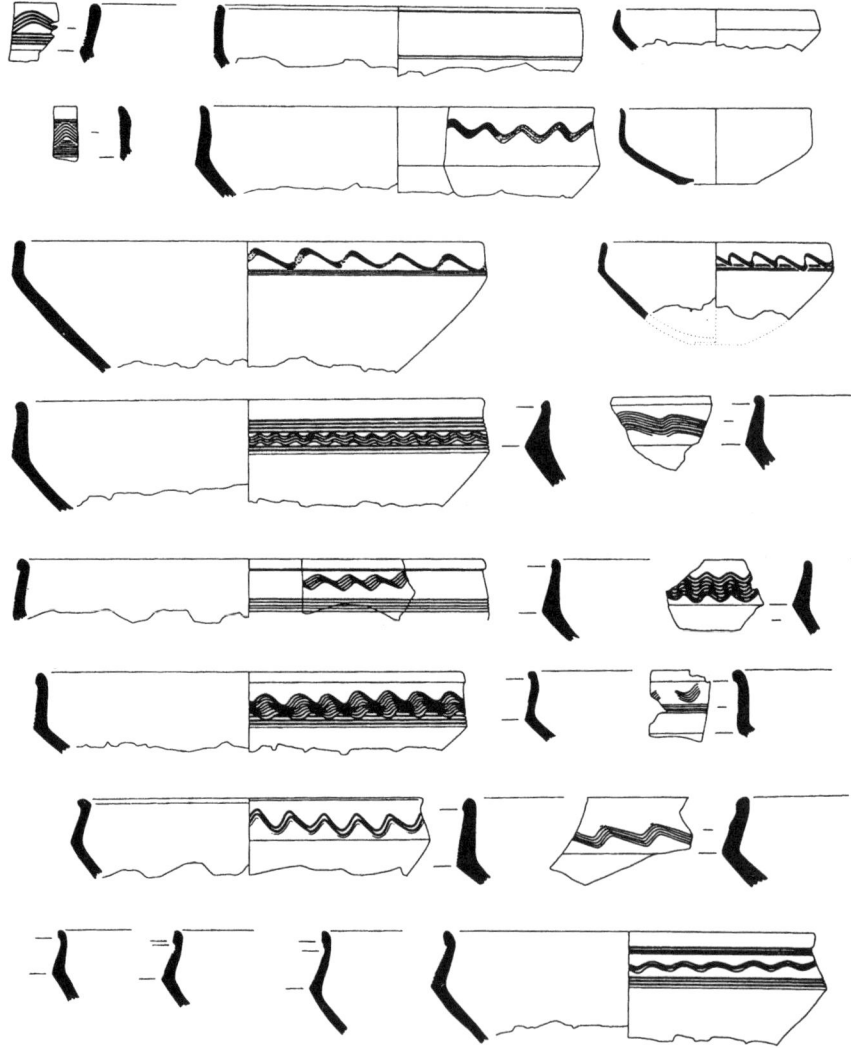

Fig. 4. Pseudo-Phocaean pottery (From Arcelin 1984, fig. 35).

1982, 54). The second area is in Provence, to the east of the Rhône where she registers a total of about 8000 fragments corresponding to about 3805 complete vessels (Arcelin 1984, 6, 85-86).

Not only has this analysis made it possible to distinguish between imported and locally-produced pottery, but it also establishes the basis on which to divide pseudo-Phocaean pottery into nine groups, each of which can be differentiated from the others on the basis of the nuances

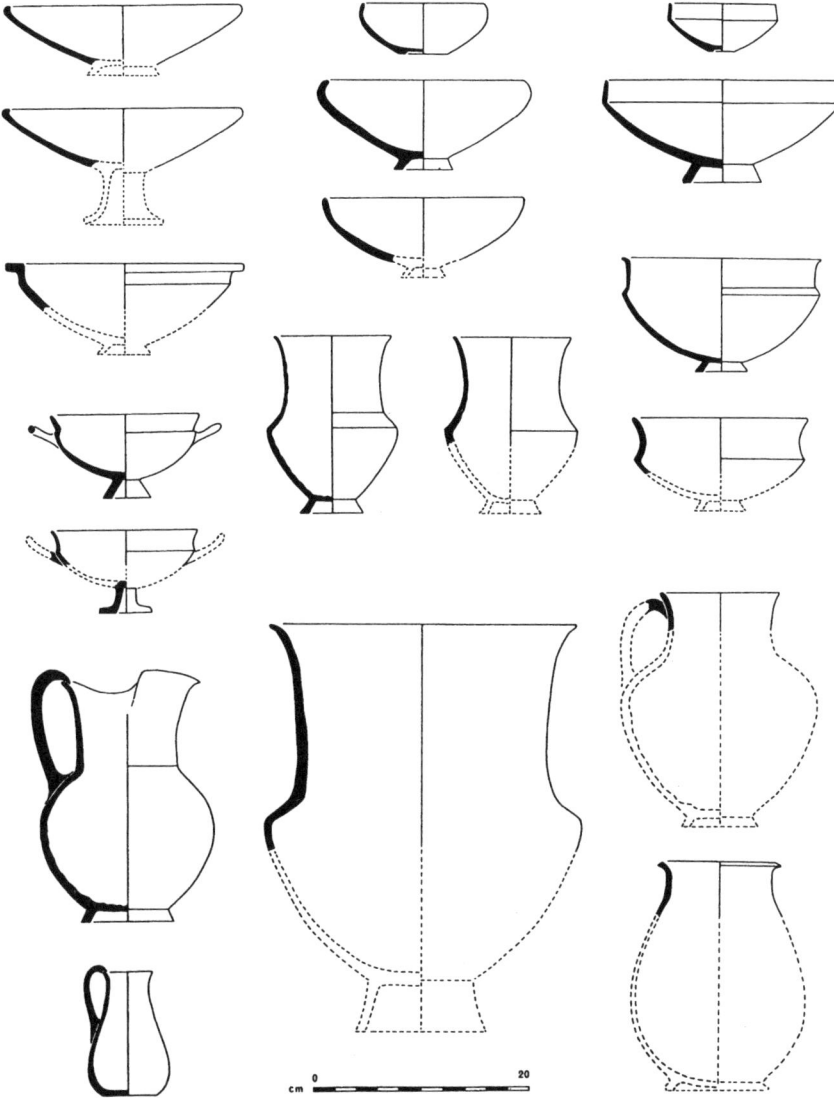

Fig. 5. The commonest shapes among pseudo-Phocaean pottery (From Arcelin 1984, fig. 3).

in the type of clay employed, production technique, shape and decoration. Arcelin-Pradelle argues that each group was produced in one or more workshops within a single area. One of these groups (Group 3) can e.g. be characterised by the fact that the ware was produced without a potter's wheel (Arcelin 1984, 6, 86, fig. 19).

The greater part of pseudo-Phocaean pottery can be dated to 525-475 BC, though production did continue, with slight modifications throughout the 4th century BC (Arcelin 1984, 145-146, fig. 22). Findspots and quantities found are given in **Figs. 1** and **2**. It is germane at this point to note that only a small number of vessels from Arcelin-Pradelle's Group 3 have been found in Marseilles and its environs compared to about 50 per cent of the total number of pseudo-Phocaean finds from the Rhône Valley; conversely 25 per cent of the few pseudo-Phocaean fragments which have been found in Central Europe, do, in fact, belong to this group (Schwab 1982, 367-372, fig. 3.2; Arcelin 1984, 85-87; Trément 1999, 110). These fragments have, moreover, been found together with Attic pottery (**Fig. 2**).

Pseudo-Ionian Pottery
Pseudo-Ionian pottery is a *painted* plain ware, which was also of local manufacture, initially produced in Marseilles proper and later in the Marseilles region and the Rhône Valley in the period 550-400 BC. Like the pseudo-Phocaean pottery it resembles an East Greek (Ionian) model, which was imported into Marseilles and its environs from around 600 BC onwards and which, moreover, principally consisted of banded drinking cups (**Figs. 6, 7** and **8**).[1]

Pseudo-Ionian pottery is differentiated from pseudo-Phocaean partly by clay-type, partly by decoration, which was painted on instead of incised. The production technique implies knowledge of painting with a brush as well as controlled firing and the use of a potter's wheel (Villard 1960, 58; Benoit 1965, 146-152).

Pseudo-Ionian pottery exists in two types according to details of shape and decoration (**Figs. 7** and **8**). The first and commonest type (A) is still very close to its East Greek model both as regards shape and decoration, the vessels usually being banded drinking cups, hence described as banded ware (**Fig. 7**). The second type (B) has liberated itself to a much greater extent from the influence of its model and become more experimental with regard to shape and decoration (**Fig. 8**). Type B is called *subgeometric*, on account of the similarity of the decorative motifs to those of the East Greek Subgeometric style, though in reality, there were probably other more directly influential sources of inspiration such as the contemporary local ware with comparable decoration probably produced at the Central European site of Mt. Lassois.[2]

THE MEDITERRANEAN AND CENTRAL EUROPE

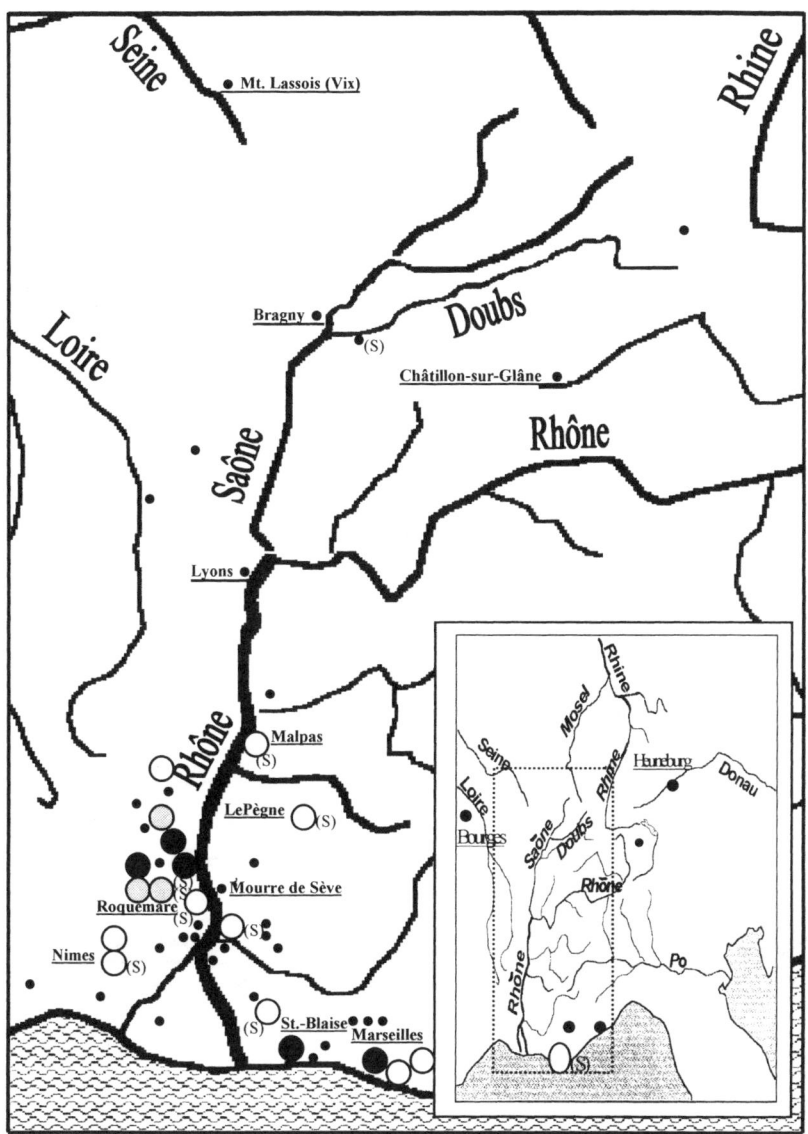

Fig. 6. Distribution of finds of pseudo-Ionian pottery from the north coast of the Golfe du Lion, the Rhône Valley and Central Europe (Based on the author's observations). More than 100 fragments found: Black circle. 30-100 fragments found: Grey circle. 5-30 fragments found: White circle. Less than 5 fragments found: Black dot. Only fragments classified according to shape are included. Since the publication of pseudo-Ionian pottery is often inadequate, this factor determines the relative difference in the enumeration in Fig.1 from that of Fig. 6. Find quantities, dominated by fragments with subgeometric decoration are marked (s).

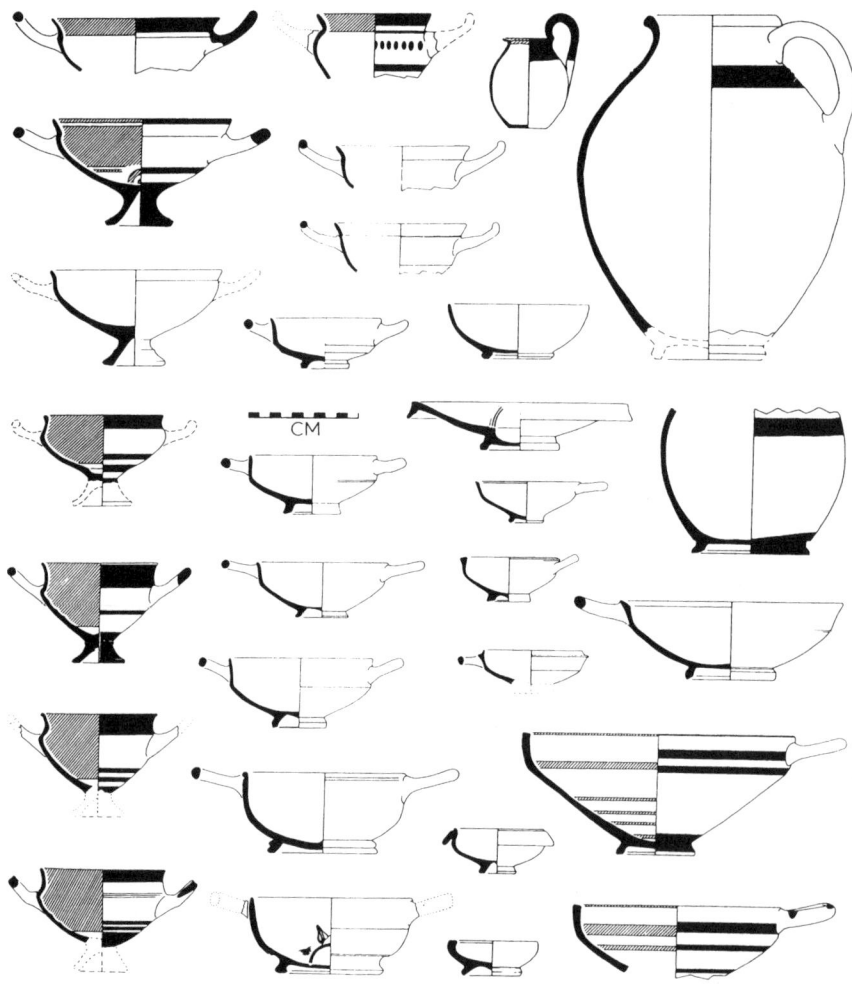

Fig. 7. Pseudo-Ionian Pottery with banded decoration (Type A) (From Py 1990, doc. 161).

There exists neither a comprehensive register nor systematic analysis of pseudo-Ionian pottery. The present study has dealt with a total of 5602 pseudo-Ionian fragments, classified according to shape, of which more than seventy-five per cent were found in St. Blaise and Marseilles: there is no doubt that the real total is much greater. Pseudo-Ionian pottery has been discovered in at least ten sites in Central Europe where it

THE MEDITERRANEAN AND CENTRAL EUROPE

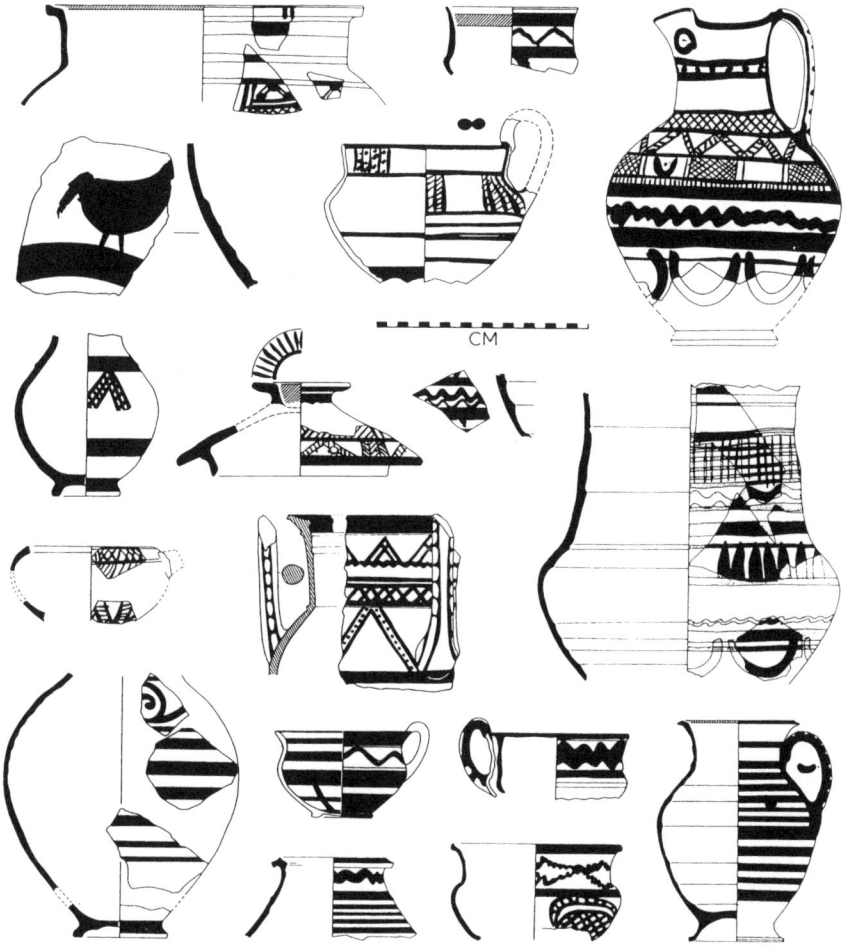

Fig. 8. Pseudo-Ionian Pottery with subgeometric decoration (Type B) (From Py 1990, doc. 162).

has included both fragments with subgeometric (type B) and banded decoration (type A) (Joffroy, 1960, 121-122; Goury, 1995). Even if pseudo-Ionian ware is included in excavation reports it is rarely classified according to shape. The same applies to the appearance in recent excavations in Languedoc of pseudo-Ionian pottery; these circumstances can, unfortunately, in no way contribute to an accurate, detailed picture of the material recovered as shown in **Figs. 3** and **6**. (Py 1990, 546-550).

155

These conditions do not yet permit a division of the material into groups to the same extent as has been possible with the pseudo-Phocaean pottery. Some individual groups of related pottery have however been documented on the basis of clay analyses confirming e.g. that the subgeometric pottery from Le Pegue, Mourre de Seve and Roquemare is interrelated. A mix of pseudo-Phocaean and pseudo-Ionian techniques is also characteristic of the region in which these three sites are to be found: the same applies to the rare examples of figure-decorated subgeometric pottery i.e type B (Lagrand 1963, 51-52, 79-82; Benoit 1964, 30-32; Benoit, 1965 169, note 47; Goury 1995, 312, 322, note 10, fig. 7.61-63).

In general it is characteristic that "open" shapes, such as bowls and cups predominate, as with pseudo-Phocaean ware, but the pseudo-Ionian pottery does however use many more Mediterranean shapes (**Fig. 3**).

It is also characteristic that there is a definite link between shape and choice of decorative motif, in that the small and "open" shapes are predominantly banded (type A) (**Fig. 7**). Subgeometric decoration is chiefly to be found on the larger and "closed" shapes, probably because these lend themselves better to the experimental subgeometric motifs (type B) (**Fig. 8**).

Pseudo-Ionian pottery dates from the middle of the 6th century BC and production continues until the 4th century BC. The majority of the banded cups appear before 475 BC, but the greater part of the subgeometric wares can probably be dated to later in the 5th century BC (Arcelin 1984, 146; Py 1990, 547, doc. 160c, 550-551, notes 175-177; Goury 1995, 310-312; **Figs.** 6 and 2).

Only small quantities of subgeometric pottery (type B) have been found in Marseilles and its environs, and this was, moreover of poor quality. In the Rhône Valley, on the other hand, subgeometric pottery is both relatively widespread and of a consistent, high quality.[3]

In Central Europe fragments have been found of both banded (type A) and subgeometric pottery, but only in modest quantities (Joffroy 1960, 121-122 **Fig. 2**).

This means that the subgeometric pottery has a pattern of distribution similar to that of the pseudo-Phocaean pottery from Arcelin-Pradelle's Group 3 (**Fig. 3**). However, as already stated, the subgeometric material generally appears later than pseudo-Phocaean pottery, and uses more of the Mediterranean shapes. This may be interpreted as a result of a more developed and firmly-established contact with Mediterranean culture.

Other Plain Wares
In addition to pseudo-Phocaean and pseudo-Ionian pottery, there were other types of plain ware which appeared in the same localities. These can also be useful when evaluating the structure of the commercial links between Central Europe and the Mediterranean countries via the Rhône Valley. This covers amphoras produced in Marseilles, Etruscan pottery and plain ware produced in Central Europe (the so called *Céramique cannelée*).

Massaliote Amphoras: Even though the definition of Massaliote amphoras is a matter of debate, the relevant finds from this study were fairly certainly produced in Marseilles (Bats et al. 1992, 464; Bertucchi 1992, 23-24). The amphoras have been found in large numbers, not only in Marseilles itself, but also in Central Europe (**Fig. 9**). This is a coarsely executed, undecorated ware, probably produced for the transport of everyday commodities from Marseilles (Benoit 1965, 182-183; Py 1990, note 208, 557; Bertucchi 1992, 185-191). These amphoras were produced in Marseilles during the period 540-100 BC, though the production appears to have peaked in the period 530-350 BC (Py 1990, 554-561, doc. 168; Gantès 1992, fig. 2; **Fig. 9**, 1-2).

As is evident from **Figs. 1** and **6**, most of the locally-produced plain ware which has been discovered was found in the lower Rhône Valley. The same can be said of the Massaliote amphoras, though in recent

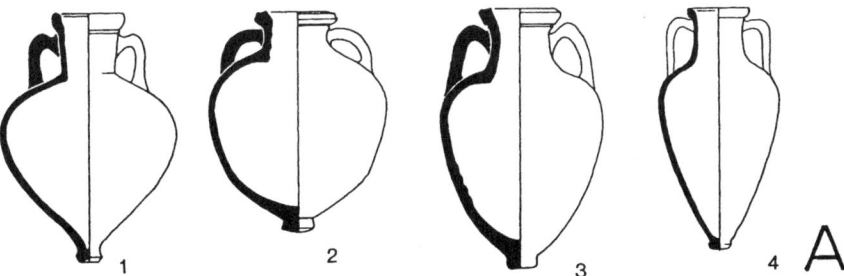

Fig. 9. Massaliote amphoras (From Py 1990, doc. 166).

years, however, these have also been found in increasing quantities in the upper Rhône Valley. Massaliote amphoras were also present in several shipwrecks discovered in the Golfe du Lion outside Marseilles (Feugere and Guillot 1986, fig. 47; Bertucchi 1992; Long, Miro and Volpe 1992, 229).

Etruscan Pottery: Etruscan amphoras began arriving in small numbers in the Rhône Valley in the mid 7th century BC of which some have been discovered still bearing traces of their contents, resinated wine. The number found in Central Europe, though, is negligible (Py 1990, 531-534; Gran-Aymerich 1992, 334-336, 339-341; Kimmig 1992, 287-288, note 18). Conversely, many have been found on the littoral region and still more among underwater finds off the coast of Southern France. It seems as though, until around 530 BC, merchant ships tended to carry more Etruscan amphoras than Massaliote ones. Thereafter the balance swings in favour of the Massaliote vessels (Gantès 1992, 172; Long, Miro and Volpe 1992, 229; Pomey and Long 1992, 189-197; Trément 1999, 109, fig. 47).

In addition, considerable quantities of Etruscan black bucchero have been discovered, dating from the same period. Most has been found in the Rhône Delta and along the coast, whereas only small quantities have been found in the Rhône Valley, and it is doubtful if any Etruscan black bucchero has been discovered in Central Europe (Py 1990, 529-531; Gran-Aymerich 1992, 340-341, note 52, 345; Gran-Aymerich 1995, 73-74, note 6; **Fig. 2**).

Céramique cannelée: This is a black Central European fine ware, which has yet to be found outside that area (**Fig. 10**). It was produced in the late Hallstatt period (540-475 BC) and demonstrates familiarity both with controlled firing and a potter's wheel. It bears a clear resemblance to metal vessels with its polished surface and profiled grooves in horizontal fields, hence its name (Lang 1970; Feugère and Guillot 1986, 172; Pare 1989, 448). It is generally thought that the model for this ware is Etruscan black bucchero, but one should bear in mind that no pottery which can be unequivocally labelled Etruscan black bucchero has yet been discovered in the relevant areas in Central Europe. This is one reason why the present author's contention is that this local ware draws equal inspiration from pseudo-Phocaean pottery, with which it shares both dating and context in the period beginning about 550 BC.[4]

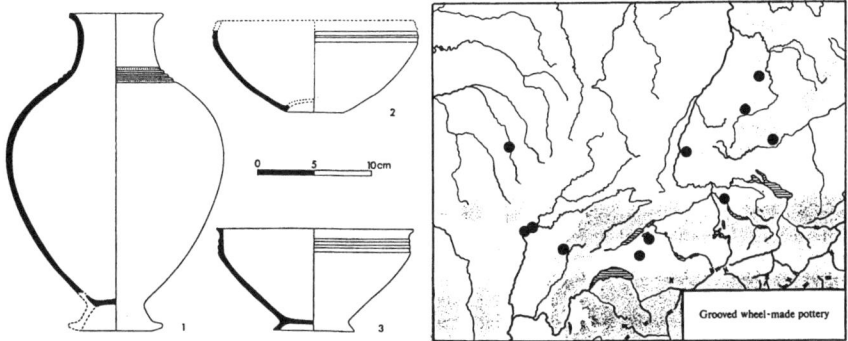

Fig. 10. *Céramique cannelée*. Findspots and the most typical shapes (From Pare 1991, fig. 1 and 2).

Imported Wares from Mediterranean Countries
In the present context there are, as mentioned, two centres in Central Europe which are relevant. Finds from the Late Hallstatt period (540-475 BC) include, from tombs, Greek, Etruscan, and occasionally, North Italian bronzes, and from dwellings, Attic pottery. From the Early La Tène period (475-400 BC) we have Etruscan and North Italian bronze vessels from tombs as well as Attic pottery in both tombs and dwellings. Moreover, these objects are usually found on the same sites (Reim 1968, 282-285; Kimmig 1981, fig. 157 and 1983, fig. 32; **Fig. 2**). Based on these series of archaeological evidence, the following discussion focuses on the organisation of the trade network linking the Mediterranean littoral to Central Europe through the Rhône Valley.

Greek Activity in Marseilles and the Rhône Valley
In Marseilles trade in Attic pottery commenced about 575 BC, i.e. shortly after its foundation (Villard 1960, 33, 43, 56-60; Gantès 1992, fig. 2-3). Beginning around 540 BC the local production of transport amphoras as well as that of pseudo-Phocaean and pseudo-Ionian plain ware largely superseded the import of such vessels from East Greece (Long, Miro and Volpe 1992, 203-206, 229; Bertucchi, Gantès and Tréziny 1995, 367-370). The pseudo-Phocaean and pseudo-Ionian plain wares were also produced in the Rhône Valley, in fact the area was to witness the gradual development of a manufacture independent of

Marseilles, something which is apparent from the distribution pattern of the finds of subgeometric and pseudo-Phocaean pottery from Group 3 (**Figs. 1** and **6**).

Judging by the distribution of the finds, this locally-produced ware was not intended for export, but for local consumption. However, the discovery in Central Europe of small quantities of these products indicates that there was direct contact between that region and the Rhône Valley (**Fig. 2**).

Since the locations in the Rhône Valley which yielded the greatest finds are strategically placed to give them direct access to the Rhône or its tributaries, it seems only logical to identify these settlements as trading points between Marseilles and Central Europe (**Figs. 1** and **6**). In such places goods from the Mediterranean destined for Central Europe by way of the Rhône may have changed hands and either underwent a further transaction through the agency of a middleman, or could be stockpiled to be shipped out later. It seems reasonable to conclude that the population in these centres numbered amongst them not only the indigenous inhabitants but also immigrants from the Mediterranean who were familiar with the necessary techniques for the production of Greek pottery.

Mourre de Seve, Roquemare and Le Pegue are examples of such strategically-located settlements in the Rhône Valley (**Figs. 1** and **6**). The contention that these places functioned as way-stations is supported in each case by the discovery of large numbers of transport amphoras, not merely of Massaliote origin, but including also locally-produced subgeometric amphoras. This argument is further strengthened by the discovery in one locality of the remains of buildings which may well have served as warehouses.[5]

In the upper Rhône Valley discoveries have been made of remarkably small quantities of pseudo-Phocaean and pseudo-Ionian pottery, as well as other finds with links to the Mediterranean. Some scholars have used this to buttress the idea that the Rhône was of no significance as a trade route between the Mediterranean world and Central Europe, a notion which still has some influence today. According to this theory the bulk of the trade between these two regions was accomplished by way of a route across the Alps (Jully 1957; Joffroy 1960, 148-153).

There may, however be other explanations for this phenomenon which do not exclude the Rhône as an important trade link. The geographical factors may have been partly responsible for rendering the area less conducive to the establishment of trading points; another factor may be that archaeological activity has been less extensive in the upper part of the Rhône Valley than in the lower. Apropos of this last observation, recent excavations in the upper Valley have, as already mentioned, turned up a significant number of Massaliote transport amphoras. The Rhône's claim to be an important trade-route is also supported by the fact that the *céramique cannelée* of Central Europe has many features in common with pseudo-Phocaean pottery.

Etruscan Activity in St.-Blaise and the Rhône Valley
Archaeological material from St. Blaise, situated to the east of the Rhône Delta near Marseilles (**Fig. 1**), calls to mind the image of a flourishing Etruscan trading point, whose activity dates back as far as the 8th century BC, i.e. long before the foundation of Marseilles itself. Commercial activity was particularly intense in the period 625-550 BC, after which it went into a gradual decline (Arcelin 1986, 82-86, note 106; Bouloumié 1992, 9-18; 63-270). This picture is repeated in other Etruscan trading points towards the Ligurian Sea, a probable example being Genoa, as well as others on the north-west littoral (Aigner-Foresti 1992). As commercial activity decreased in St. Blaise and the other trading centres towards the Golfe du Lion and the Ligurian Sea, Etruscan trade increased in the Po Valley, from which contact with Central Europe had already been established by way of the Alps (Dobesch 1992, 169-171; Pauli 1992, 179-182; Nilsson 1999, 15-16, 19).

The declining traffic in goods integral to Etruscan activity following the foundation of Marseilles has often been taken as evidence that the market of this area including connections to Central Europe was insufficient to support both Etruscan and Greek commercial interests. However the change in the commercial activity was only gradual throughout the whole of the 6th century BC as may be concluded from the following evidence:

In St. Blaise most of the transport amphoras in use were still Etruscan until about 500 BC. Moreover, more pseudo-Phocaean and pseudo-Ionian pottery has been discovered in St. Blaise than in Marseilles (Arcelin 1986, note 106; Py 1990, 531, doc. 152; Bertucchi 1992, 212; Bouloumié 1992, 101-172; Trément 1999, 109, fig. 47, 110).

In Marseilles discoveries from the 1980s have shown that the occurrence of Etruscan pottery increased markedly compared with the amount of imported Greek ware in the period 575-550 BC (Gantès 1992). As already stated, it was not until after 550 BC that Marseilles' own production came to predominate.

This picture, which is also confirmed by the ratio of Etruscan to Massaliote amphoras from wrecks of merchant ships in the Golfe du Lion, shows no radical changes, possibly as a result of a conflict of interests between Marseilles and St. Blaise (cf. above).

The location of St. Blaise makes it probable that the Etruscan commercial goods were distributed along the Rhône as a link to Central Europe. Since however, as mentioned above, the finds of Etruscan pottery are very limited in the Rhône Valley, and almost non-existent in Central Europe, it seems unlikely that these Etruscan goods were distributed by merchants of Etruscan origin. On the other hand they may have been distributed through the trade network organized by local merchants, which may well have been in operation in the Rhône Valley before the foundation of Marseilles, but able to serve Etruscan interests in St. Blaise in the 7th century BC. At all events the peoples of the Rhône Valley must have been familiar with Etruscan pottery since at least ten pseudo-Phocaean copies of traditional Etruscan *kantharoi* have been found, evenly distributed throughout the Rhône Valley in St. Blaise, Le Castelet, St. Paul-Trois-Château, Castelan d'Istres (2 pieces), Mourre de Sève (2-3 pieces), La Roche du Comps, La Marduel and Lyons.

Trade: the relative significance of the various imported goods
When evaluating the commercial activity between the Mediterranean countries and Central Europe, the focus of the classical archaeologist's research has traditionally been on vessels and dress accessories in bronze as well as on Attic pottery.

However, Attic pottery was not imported into Central Europe before the late Hallstatt Period, i.e. after 540 BC, and, apparently, only for use in living quarters as, barring a single exception, no Attic pottery has been found in tombs before the La Téne Period (Joffroy 1954, 31). Furthermore two rare Attic red-figure cups found in the Kleinaspergle "Princely Tomb" at Hohenasperg in the Rhine district, were finished in metal foil in the Etruscan tradition before they were placed in the tomb together with a bronze situla from Este and an Etruscan bronze "Schnabelkanne" (Kimmig, 1981, fig. 155).

This could mean that Attic pottery was not regarded as a status symbol and seeing that the actual quantity discovered is limited, Attic pottery can hardly have been a particularly important commodity (Nilsson 1999). The few bronze show-pieces recovered probably had a social significance rather than being acquired as items of trade, whereas the simpler bronze ware seems to have enjoyed a more widespread distribution and thus was probably a more important commodity (Kimmig 1992, 293-320, 324). At all events bronzes have not been found to an extent which would make it reasonable to suggest that their distribution was dictated by commercial rather than cultural demands.

As in our own day, what really kept the economy afloat must necessarily have been the production and retail of ordinary everyday commodities (Bertucchi 1992, 185-191; Trément 1999, 119-120). Consistent with this is the fact that excavation of most of the sites in the Rhône Valley and Central Europe has yielded large numbers of Massaliote amphoras, in which such goods would have been transported; in addition, there would probably have been containers/packaging made of more perishable material such as leather.

It has often been the view that commercial activity in Marseilles slackened after 475 BC, since imports of Attic pottery decreased. Supported by an equal increase in imported Attic pottery in Spina and Adria on the Adriatic coast this theory suggested a general decline in Marseilles due to Etruscan competition (Reim 1968, fig. 2; Shefton 1994, 64).

It is thus important to state that the production of amphoras in Marseilles during this period however shows no decline, which argues against the idea of any slump in trade (Bertucchi 1992, 211-213). Quite the reverse obtained, in fact: production continued at an unchanged intensity until the middle of the 4th century BC, which means that commercial activity experienced no setbacks during this period. One is therefore compelled to look elsewhere for factors which would explain the decrease which now, in the light of recent finds from the 5th century BC, appears less severe than was previously supposed.

Conclusions
It is possible to draw the following conclusions concerning the trade links between the Mediterranean countries and Central Europe in the 6th and 5th centuries BC: The distribution of goods leaving Marseilles

and Saint-Blaise was accomplished, at least partially, through a locally-organised trade network already in existence in the Rhône Valley.

The local Greek-inspired plain wares, produced independently of Marseilles, which has been found in Central Europe indicate that the local merchants from the Rhône Valley trade network had direct commercial links with settlements in Central Europe as the plain wares in question would not appear as prime commercial commodities.

A focus on the distribution of amphoras whose original primary purpose was the transport of everyday necessities presents a different perspective from the one derived from examination of more prestigious items such as Attic pottery and bronze ware.

The distribution patterns of Etruscan and Massaliote amphoras respectively demonstrate that the commercial activities in Saint Blaise were not seriously affected by the foundation of Marseilles in 600 BC. The most important conclusion, however, is that the intensity of the trading activities in Marseilles via the Rhône Valley remained constant throughout the period 540-400 BC.

Lone Leegaard
Manøgade 7 (2.tv.)
DK-2100 Copenhagen Ø

Tel: 39 27 27 76

E-mail: lonelee@worldonline.dk

NOTES

This article was translated into English by Neil Stanford.

1. Vallet and Villard 1955, 15-30; Villard 1960, 43; Lagrand 1961, 57; Bats et al. 1992; 461-463; Bouloumié 1992, 101-172; Long, Miro and Volpe 1992, 203-206.

2. Jully 1957; Hatt, Perraud and Lagrand 1960, fig. 5; Benoit 1964, 47; Villard 1960, 62; Benoit 1965, 170-174; Lagrand 1963, 76-78; Py 1990, 546, notes 151-152, 548, doc. 162.

3. Benoit 1964, 41; Benoit 1965, 169; Py 1990, 546-550; Goury 1995, 312-313, 321-323.

4. Joffroy 1960, 119, 143, pls. 64-66; Benoit 1965, pl. 23; Schwab 1982, 457; Unz 1985, 324, fig. 505; Long, Miro and Volpe 1992, 30-31; Gran-Aymerich 1995, 73, note 6.

5. Perraud 1955; Jully 1957, 50-56; Lagrand 1963, 79-82; Benoit 1964; Benoit 1965, 177-180; Goury 1995, 321-322; Trément 1999, 119.

BIBLIOGRAPHY

Aigner-Foresti, L. 1992
Etrusker im Land der Ligurer: Merkmale und Bedeutung ihrer Anwesenheit, in: L. Aigner-Foresti (ed.), *Etrusker nördlich von Etrurien* (Wien 1989), 103-111.

Arcelin, P. 1986
Le territoire de Marseille grecque dans son contexte indigène, *Études Massaliètes* 1, 43-104.

Arcelin-Pradelle, Ch., Dedet, B. & Py, M. 1982
La céramique grise monochrome en Languedoc oriental, *RANarb* 15, 19-67.

Arcelin-Pradelle, Ch. 1984
La céramique grise monochrome en Provence. *RANarb suppl.* 10.

Bats, M. et al. 1992
Resumé des discussions, *Études Massaliètes* 3, 460-476.

Benoit, F. 1964
La céramique peinte de Roquemare à l'époque grecque, *CahRhodBord* 11, 30-61.

Benoit, F. 1965
Recherches sur l'hellénisation du Midi de la Gaule. Aix-en Provence.

Bertucchi, G. 1992
Les amphores et le vin de Marseille. *RANarb* suppl. 25.

Bertucchi, G., Gantès, L. F. & Tréziny, H. 1995
Un atelier de coupes ioniennes à Marseille, *Études Massaliètes* 4, 367-370.

Bouloumié, B. 1992
Saint-Blaise (Fouilles de H.Rolland). Aix-en-Provence.

Dobesch, G. 1992
Die Kelten als Nachbarn der Etrusker in Norditalien, in: L. Aigner-Foresti (ed.), *Etrusker nördlich von Etrurien* (Wien 1989), 161-178.

Etrusker nördlich von Etrurien
L. Aigner- Foresti (ed.), *Akten des Symposions von Wien- Schloss Neuwaldegg 1989.* (Öst. Akad .Wiss. phil.-hist. Sitzungsberichte 589 I-II),Wien 1992

Feugère, M. & Guillot, A. 1986
Fouilles de Bragny 1. Les petits objets dans leur contexte du Hallstatt final, *RAE* 37, 159-221.

Gantès, L.-F. 1992
L'apport des fouilles récentes à l'étude quantitative de l'économie massaliètes, *Études Massaliètes* 3, 171-178.

Goury, D. 1995
Les vases pseudo-ioniens des vallées de la Cèze et de la Tave (Gard), *Études Massaliètes* 4, 309-324.

Gran-Aymerich, J. 1992
Les materieaux étrusques hors d'Etrurie, in: L. Aigner-Foresti (ed.), *Etrusker nördlich von Etrurien* (Wien 1989), 329-359.

Gran-Aymerich, J. 1995
Griechische Vasen und etruskische Bronzen aus Bourges in ihrem archäologischen und historischen Kontext, in: I. Wehgartner & H. Zöller, *Luxusgeschirr keltischer Fürsten* (exhibition Mainfränkischen Museums Würzburg), 71-74.

Hatt, J.J., Perraud, A. & Lagrand, Ch. 1961
Le Pegue, habitat hallstattien et comptoir ionien en haute Provence, in *Atti del settimo congresso internazionale di archeologia classica* 3, Roma, 177-186.

Jacobsthal, P. & E. Neuffer, 1933
Gallia Graeca. Recherches sur l'hellenisme de la Provence. (Préhistoire 2.1). Paris.

Joffroy, R. 1954
Le trésor de Vix (Côte-d'or). Paris.

Joffroy, R. 1960
L'oppidum de Vix et la civilisation hallstattienne finale. Paris.

Jully, J. J. 1957
À propos de la céramique de la colline St.Marcel (Le Pegue, Drôme), *CahRhodBord* 4, 49-56.

Justinus, *Epitome*.

Kimmig, W. 1992
Etruskischer und griechischer Import im Spiegel westhallstättischer Fürstengräber, in: L. Aigner-Foresti (ed.), *Etrusker nördlich von Etrurien* (Wien 1989), 281-328.

Kimmig, W. 1983
Die griechische Kolonisation im westlichen Mittelmeergebiet und ihre Wirkung auf die Landschaften des westlichen Mitteleuropa, *JbZMusMainz* 30, 5-78.

Kimmig, W. 1981
in: Bittel, K. (ed.), *Die Kelten in Baden-Württemberg*, Stuttgart, 248-281.

Lagrand, Ch. 1963
La céramique "pseudo-ionienne" dans la Vallée du Rhône, *CahRhodBord* 10, 37-82.

Lamb, W. 1932
Grey Wares from Lesbos, *JHS* 52, 1-12.

Lang, A. 1970
Geriefte Drehscheibenkeramik der Heuneburg 1950-70 und verwandte Gruppen. Berlin.

Long, L., Miro, J. & Volpe, G. 1992
Les épaves archaïques de la pointe Lequin, *Études Massaliètes* 3, 199-234.

Maffre, J.-J. 1995
Funde griechischer Keramik in Ostfrankreich, in: I. Wehgartner & H. Zöller, *Luxusgeschirr keltischer Fürsten* (exhibition Mainfränkischen Museums Würzburg), 64-65.

Nilsson, A. 1999
The Function and Reception of Attic Figured Pottery. Spina, a Case Study, *ARID* 26, 7-23.

Pare, C. F. E. 1991
Fürstensitze, Celts and the Mediterranean World: Developments in the West Hallstatt-Culture in the 6th and 5th cent.. B.C., *ProcPrHistSoc* 57.2, 183-202

Pauli, L. 1992
Die historische Entwicklung im Gebiet der Golaseccakultur, in: L. Aigner-Foresti (ed.), *Etrusker nördlich von Etrurien* (Wien 1989), 179-196.

Perraud, A. 1955
Le Pegue, préface de Marseille? Paris.

Pomey, P. & Long, L. 1992
Les premiers échanges maritimes du Midi de la Gaule du VIe au IIIe s.av. J.-C., *Études Massaliètes* 3, 189-198.

Py, M. 1990
Culture, économie et société protohistoriques dans la région nimoise (*CEFR* 131.2). Rome.

Scotto, R. F. 1985
La céramique grise à décor ondé de Montmorot (Jura), *RAE suppl.* 6, 45-51.

Shefton, B. B. 1994
Massalia and Colonization in the North-Western Mediterranean, in: *The Archaeology of Greek Colonization*, Oxford, 61-86.

Schwab, H. 1982
Pseudophokäische und phokäische Keramik in Châtillon-sur-Glâne, *Archäologisches Korrespondenzenblatt* 12, 363-372.

Trément, F. 1999
Archéologie d'un paysage. Les étangs de Saint-Blaise (Bouches-du-Rhône). Paris.

Unz, C. 1985
Typologie, *DenkmPflBadWürt* 14, 89-95.

Vallet, G. & Villard, F. 1955
Megara Hyblaea V. Lampes du VIIe siècle et chronologie des coupes ioniennes, *MEFRA* 67, 7-34.

Villard, F. 1960
La céramique grecque de Marseille (VIe-IVe siècle). Paris.

CYPRIOT TRANSPORT AMPHORAE IN THE ARCHAIC AND CLASSICAL PERIOD

KRISTINA WINTHER JACOBSEN

Petrographic and neutron activation analyses of the transport amphorae with horizontal handles excavated at Tell Keisan in Israel indicate that this type of amphorae were produced in South Eastern Cyprus, in the area around Salamis (Courtois 1980, 358-360; Gunnweg & Perlmann 1991, 596-597). Fragments of the same type of amphorae are common in the excavation of the Danish Archaeological Expedition at Panayia Ematousa, Aradippou, Cyprus. The diagnostic material comprises mainly fragments of the distinctive, large, broken-off handles and a few rims and bases. The fabrics at Panayia Ematousa can be separated into a coarse and a medium-coarse group. The coarse fabric (I) is made of gritty clay ranging from very pale brown to sometimes reddish yellow, with numerous tiny black, white, red and beige particles. A related group of fragments (Ia) is made of a very characteristic gritty clay ranging from very pale brown to pale yellow with numerous tiny greyish black grits, a few white particles, and distinctive beige limestone reaction rims. At least two amphorae of the same type found at Marion, Cyprus, dated the end of the Cypro-Archaic period II, are made of fabric Ia (Gjerstad 1935, 416 no. 80:1, 450 no. 97:4). The fragments of fabric Ia from Panayia Ematousa belong to amphorae with handles with a round section, which is considered an early feature as opposed to an oval section (Gjerstad 1960, 120-121 fig. 15). The medium coarse fabric (II) ranges from very pale brown to reddish yellow, often with a different colour at the core, and has tiny to small black, grey, brown, red, and white particles. Generally the handles with an oval section are made of the less coarse fabric II, indicating that the refinement of the shape was accompanied by a refinement of the fabric. Both fabrics are well known in Cyprus from local coarse ware and local imitations of Koan transport amphorae.

The evolution of the amphora type
The earliest transport amphorae were developed in Palestine already in the Late Bronze Age (Grace 1979a, 8 fig. 12; Sagona 1982, 73). These amphorae were small, in the shape of a bag and had two small ear-handles, and this basic shape continued to be used in the Middle East. The development of the typical Greek transport amphora began in the 8th century BC and one of the first types was the so-called SOS amphora from Attica, developed from the Late Geometric amphora with vertical handles (Johnston & Jones 1978, 132). Although it may not have been strictly developed for transport, this type of amphora was soon exported all over the Mediterranean, and the shape became the prototype for countless variations of transport amphorae in the central and western area of the Mediterranean, from modern Turkey to Spain. Through time the transport amphorae developed a longer and more slender body, and in most cases the foot became pointed, but the basic shape continued to be the same until the ceramic containers were replaced by wooden casks.

In Cyprus a different and unique shape was developed, the transport amphorae with horizontal handles (**Plate 11**). This type of amphora shares several of the characteristics of the Greek shape, but the overall impression is very different, because of the large loop or basket-handles rising high above the rim in a transversal position opposite to that of the longitudinal axe of the greek amphora.

A Cypriot preference for horizontal handles demonstrated in the small white painted and bichrome amphorae from the Early Iron Age (**Fig. 1**) seems to have influenced the development of the local transport amphora. This preference seems to have influenced the development of the local transport amphora. An impressive series of 34 amphorae from one of the royal tombs at Salamis, dated around 700 BC are the oldest identified examples (Karageorghis 1973, 52). The first fully-developed transport amphora with horizontal handles has a large biconical body and a small flat, shaved base (**Plate 11**). The center of gravity is around the middle of the belly, where traces of a cord tied around the body while the vessel was drying is frequently seen. The neck is very short with a high, everted rim, but most distinctive are the powerful horizontal handles rising from the shoulder high above the rim. The average height of the earliest transport amphorae with horizontal handles is 65 cm, but they range from 45 to 85 cm. The shape underwent morpho-

Fig. 1. Small Cypriot Iron Age white painted and bichrome amphorae (Department of Antiquities inv.no. 646-647, National Museum, Copenhagen)

logical changes, and during the Classical period the amphorae became taller and slimmer. The biconical profile gave way to a more elegant, ovoid profile, but the high horizontal handles remained an unmistakable feature (**Fig. 2a-b**).

Another Danish excavation at Tell Sukas in Syria has excavated a large number of transport amphorae with horizontal handles, and the majority of the material, published by M.-L. Buhl, is believed to be a local production (Buhl 1983, 15-23). The transport amphorae with horizontal handles produced in the Levant developed into a carrot-shaped, neckless container, the so-called torpedo shape (**Fig. 2b**). Since the bag-shape and the missing neck is typical in the Levant this type of amphorae is only recognisable from the handles. Because of the missing neck these were generally much smaller than those on the Cypriot amphorae (Humbert 1991, 587-590).

Though some fragments of amphorae with horizontal handles seem to have wheelmarks, research into near contemporary production methods in Cyprus indicates that vessels of this size were built up on a

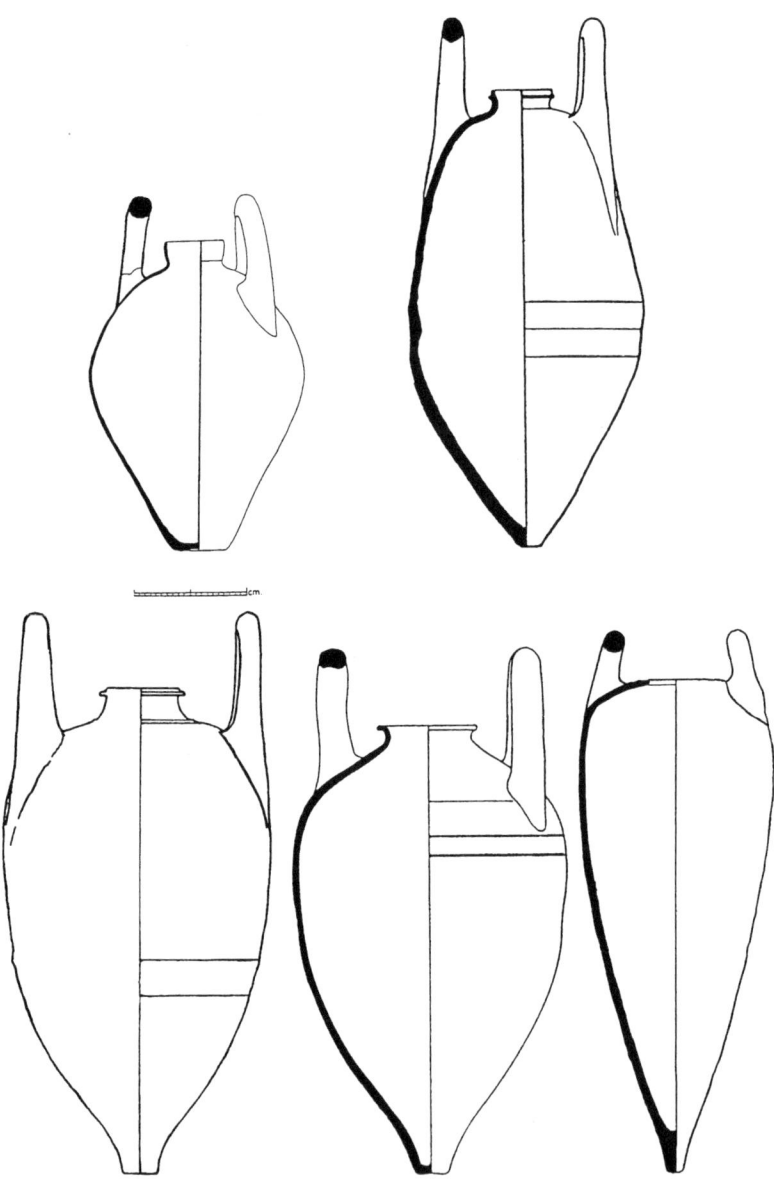

Fig. 2a. Profiles of Cypriot transport amphorae from Salamis (from Karageorghis 1973, 54, pl. 222) and from the Swedish Cyprus Expeditions (from Gjerstad 1948, fig. 57:23) (drawing by K.W. Jacobsen).

Fig. 2b. Profiles of Cypriot transport amphorae from the Swedish Cyprus Expeditions (Gjerstad 1948, fig. 62:10), from Amathos (Marchetti 1978, 947 fig. 15), and from Salamis (Karageorghis 1973, 186, pl. 287:1108) (drawing by K.W. Jacobsen).

slow turntable (London 1990). This is supported by the shaved surfaces where superfluous clay has been scraped off, and by the impressions of the cord tied around the belly to prevent the walls from collapsing while drying. All Cypriot clays contain montmorillonite, which enhances the plasticity of the clay and gives more strength to the prefired vessel, allowing the shaving of the walls (Hemsley 1991, 215). Some of the amphorae seem to have internal wheelmarks, and maybe the two halfs were made separately on a fast turntable, and the joint bound with cloth (Humbert 1991, 576).

Oil or Wine?
The contents of the amphorae with horizontal handles is not known. A painted inscription on an amphora of this type from one of the royal tombs at Salamis dated to around 600 BC, indicates that it contained oil (Karageorghis 1967, 38 no. 101, pl. XLI, CXXVI; Masson 1967, 132-133). Graffiti consisting of the letters *alef* and *lamed* on the shoulders of amphorae from Tell Keisan in Israel are interpreted by E. Puech as a Phoenician abbreviation of the contents and the provenance: *alef* for oil, *lamed* for *lšy* (Alasia) (Puech 1980, 301-303, pl. 91, 137). In the Levant, according to J.F. Salles (1991, 230), the amphorae with horizontal handles are only found in areas with no noticeable local oil-production, which speaks in favour of oil as the main contents. To put it in another way, in areas where they grow wine but have to import oil, foreign transport amphorae are most likely to have contained oil. However it is impossible to divide areas into producers and receivers exclusively, since oil was produced almost everywhere, but in many different qualities and for a variety of purposes: for example, food, body-care and lighting.

On a Cypro-Phoenician bronze bowl in the British Museum dated to the 6th century BC an amphora with horizontal handles is shown hanging from a beam carried by two men to a banquet (Markoe 1985, 57, 174-175, no. Cy5). A small terracotta figurine in a private collection in Cyprus depicts the same scene (Calvet 1986, 506 fig. 2b). The banquet scene suggests that the amphora contained wine for the party, although it might have contained scented oil for perfuming the guests (Salles 1980, 137).

Two of the fragments from Panayia Ematousa have a thick pinkish "slip" on the inside, which has also been observed in amphorae found in the Levant (Jacobsen 1998, 351 no. 93). Humbert suggested that it may be a deposit of the lime added to wine in order to prevent it from fermenting (Humbert 1991, 577).

The evidence seems therefore to suggest that the horizontal-handled amphorae carried both wine and oil. This possibility has been demonstrated before, when analyses of the contents of the Greco-Italic amphorae from the shipwreck at Capistello near Lipari proved that they carried both wine and oil (Frey 1978, 289). Cyprus has produced both wine and oil since prehistoric times, and the Roman geographer Strabo (14.6.5) praised the island for the high quality of both products.

It is important to look at the socio-economic context and the agricultural setting into which the transport amphorae belong and analyse the mechanisms by which the agricultural surplus was redistributed or exported.

The earliest evidence for olive oil production in Cyprus are the pressing installations in sanctuaries from around 1300 BC (Hadjisavvas 1992, 21-25). The sanctuaries were probably involved in the organisation of the trade too. The large number of imported objects in the Bronze Age sanctuary at Kalavassos-Agios Dimitrios, which also included an oil-pressing installation suggest international exchange, i.e. a production directed at export (Hadjisavvas 1992, 83). Luxury objects from Greek mainland santuaries have also been explained through involvement in long-distance trade (Strøm 1992, 60).

The centralised production of oil would very likely imply an equally centralised production of amphorae, although no traces of kilns have been recorded in the vicinity of the Cypriot sanctuaries so far. The producers probably moved the olives to the sanctuaries, where they were pressed and poured into amphorae. From the sanctuaries the oil would then be shipped to foreign receivers. The trade would be, at least to some extent, organised as a royal tribute, which was an important factor in the commercial relations in the eastern Mediterranean, especially under Achaemenid domination (Salles 1991, 216-217).

Similarly although in another social context, that of private enterprises, a centralised production of amphorae has been suggested at Chersonessos in the southern Crimea. Here surveys have revealed many isolated farmhouses with local pressing installations for wine, but only one amphora production site, situated near the coast, outside the city wall (Whitbread 1995, 13-19 with refs.). The Panayia Ematousa excavations may offer new evidence on the organisation of the oil production in Cyprus in Pre-Roman times. Counter-weights and a small press bed, all re-used as building material, point to a local production of oil (Sørensen 1996, 139). Unfortunately the secondary find contexts prevent a fixed date, but the finds may suggest that people around the island were supplied by local small-scale producers while the sancturies produced oil for overseas customers. The amphorae with horizontal handles were used locally, as documented by the many finds all over Cyprus. S. Hadjisavvas (1992, 117) has drawn attention to the general importance of the oil export, as indicated by the number of amphorae with horizontal handles in Archaic and Classical tombs, suggesting a large-scale production. This is supported in the Hellenistic period by the size of the pressing installations and the technological improvements.

As to viticulture, pips from cultivated grapes turn up in Cyprus in the 2nd millennium BC, but there is no direct evidence for an ancient wine production, unless oil presses could have been used for both purposes (Karageorghis 1993, 31-32). Oil and wine production require somewhat similar pressing installations, and it can be difficult to distinguish between the two without analyses of the organic material (Brun 1993, 511, 517). However, ancient depictions of wine presses, e.g. vase paintings, indicate that less permanent wooden structures that would have left no archaeological traces were preferred for the pressing of grapes until the end of the 5th century BC (Isager & Skydsgaard 1992, 57).

Distribution of Cypriote amphorae
Outside Cyprus, transport amphorae with horizontal handles have so far been found in most neighbouring areas in the Eastern Mediterranean: South Western Turkey (Alpözen 1995, 70-71), Western Rhodes (Jacopi 1929, nos. 77, 121 og 129, pl. IV; Jacopi 1931, nos. 131, 142, 149, 158-160, 210, pl. VIII), the Levantine coast, and Egypt (Stern 1973, 111; Salles 1980, 139; Sagonas 1982, 106-107; Ben-Tor 1986, 44 fig. 4:16-17), though never in very large numbers and never further away (**Fig. 3**).

The distribution map reveals a considerable concentration in the area between Tell Keisan in the north and Tell Fara in the south, an area of Israel which did not produce olive oil in Antiquity (Salles 1991, 228). The petrographic and neutron activation analyses of the transport amphorae with horizontal handles linking Israel with South Eastern Cyprus (Courtois 1980, 358-360; Gunnweg & Perlmann 1991, 596-597)

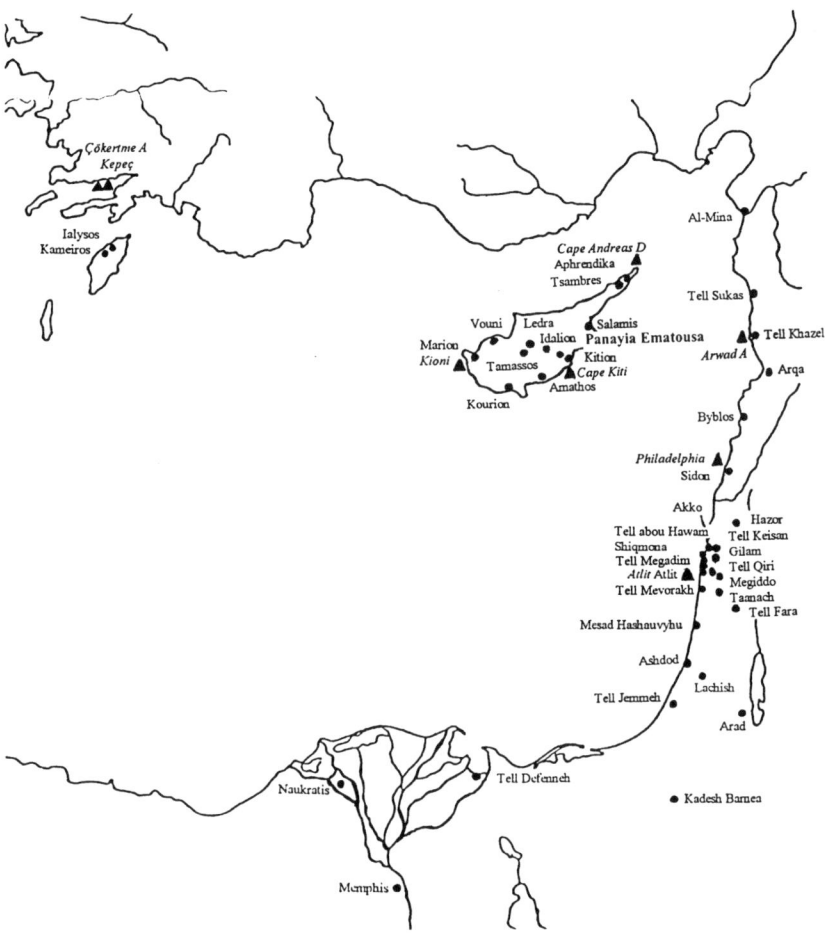

Fig. 3. Distribution of transport amphorae with horizontal handles. *Italics*: amphorae found at sea (K.W. Jacobsen).

testify to trade patterns different from those prevailing in the Roman period. The fine wares of the Roman period show close relations between South Eastern Cyprus and North Western Syria on one hand, while on the other hand Western Cyprus had close relations with Israel, Egypt, and Cilicia (Lund 1999, 12).

Three underwater contexts, two archaic wrecks found off the Turkish coast and one off Arwad in Syria, are of special interest because two of them appear to be entire shipments of amphorae with horizontal handles (Parker 1992, 60 no. 58, 148 no. 324; Alpözen 1995, 71). The last wreck, found off Kepeç, included amphorae with horizontal handles, but not enough information is available to determine if it was an entire shipment (Parker 1992, 226 no. 542; Alpözen 1995, 70). Regrettably, the three wrecks were in very poor condition, making it difficult to judge the provenances of the ships carrying the Cypriote amphorae.

In 1996 a shoulder fragment was found in an archaic well by the temple of Athena at Milet (Niemeier 1999, 389-92) and hopefully more fragments will be identified in Turkey in the future.

In Rhodes amphorae with horizontal handles were used for infant burials in the possibly Phoenician settlements at Kameiros and Ialysos (Jacopi 1929 and 1931). The fact that the Cypriot amphorae in Rhodes only seem to appear in Phoenician settlements suggests a close connection between the distribution of the Cypriot amphorae and the Phoenician traders.

The end of productions
Some time, probably during the Hellenistic period, the production of amphorae with horizontal handles came to an end. The amphora fragments from the pressing installation in Nicosia and from Amathos show that the amphorae by then had evolved into an ovoid amphora-type with a small, hollow, pointed toe (Hadjisavvas 1992, 31 no. 6, fig. 55; Marchetti 1978, 446, fig. 15 and 18), the transition to the more common shape being close by. In Cyprus the production of transport amphorae of the Greek type with vertical handles began at the end of the 4th century BC. Amphora production centres have been identified at Kourion, Paphos, Salamis, and probably also at Kition (Grace 1979b, 178-188). This new amphora type is believed to have carried wine (Calvet 1986, 505), and it is in fact an imitation of the slightly earlier amphora type from the island of Chios, in antiquity famous for its excellent wine (Grace 1979b, 183).

It is generally agreed that transport amphorae with horizontal handles were not produced in Cyprus after the 4th century BC, while amphorae of the same tradition continued to be produced in the Levant down into the Hellenistic period (Humbert 1991, 586-588).

Fig. 4. Two Cypriot transport amphorae in room VII, Panayia Ematousa, Cyprus.

At Panayia Ematousa fragments of two amphorae with horizontal handles have been found almost at the bottom of the pit-like Room VII, together with fine ware dated to the 3rd and 2nd centuries BC, and it seems unlikely that the amphorae would be intrusive (**Fig. 4-5**). The fragments are made of the medium-coarse clay, typical of this type of amphorae and other Cypriot coarse wares. One of the fragments is very similar to Gjerstad's last types VI-VII, dated to 475-325 BC (Gjerstad 1960, 120-121 fig. 15), but the section of the handles from Panayia Ematousa is ovoid and probably flat where the handles bend. This feature is typical of amphorae with horizontal handles produced in the Levant (Humbert 1991, 588 type K). But, as the clays seem to be Cypriot, it may also be a late feature of the Cypriot production. It is, of course,

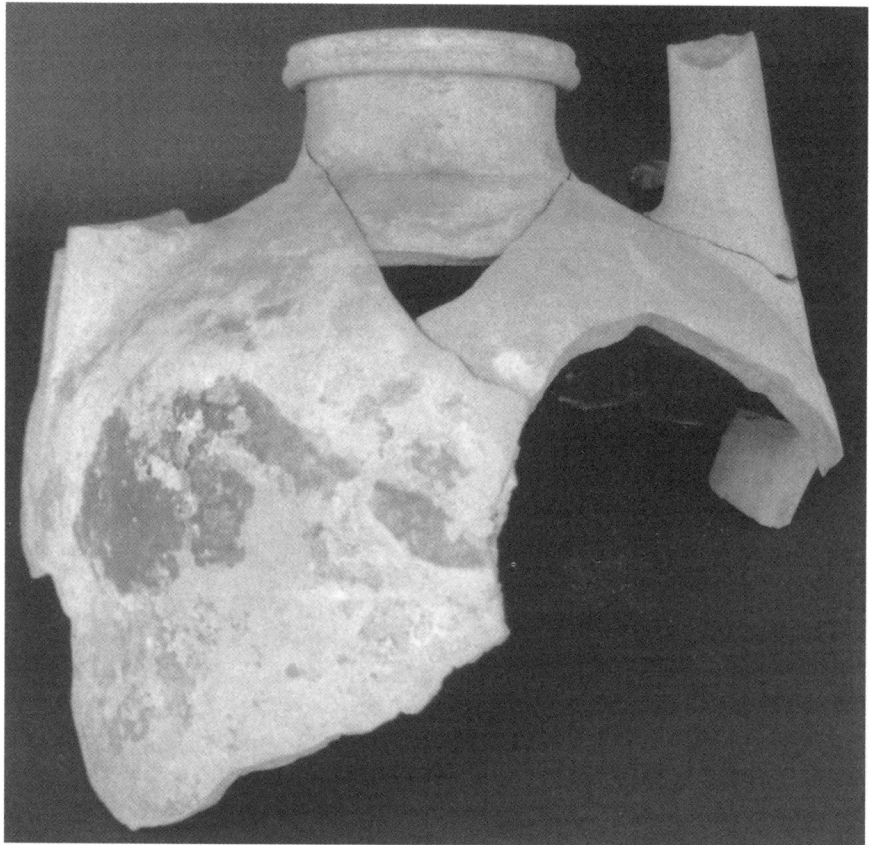

Fig. 5. One of the Cypriot transport amphorae from room VII, Panayia Ematousa, Cyprus.

possible that the amphorae in question were old, having been reused over a long period of time. Sale of empty transport amphorae has been documented in Athens (Amyx 1958, 174-175), and Humbert suggested that the amphorae with horizontal handles were reused in the Levant because of their solidness (Humbert 1991, 576). Repairs are, in fact, rare, but they do appear, as is seen on a fragment from Panayia Ematousa and on an amphora from Salamis (Jacobsen 1998, 351 no. 90; Karageorghis 1970, 140 tomb 59 no. 6, pl. CXL:6) (**Fig. 6**). However, although the Cypriot amphorae with horizontal handles are thick-walled and the handles are very powerful, they tend to be very fragile in the area where the handles are attached to the body. The heavy fabric and thick walls made the amphorae very heavy, and it is the weight which

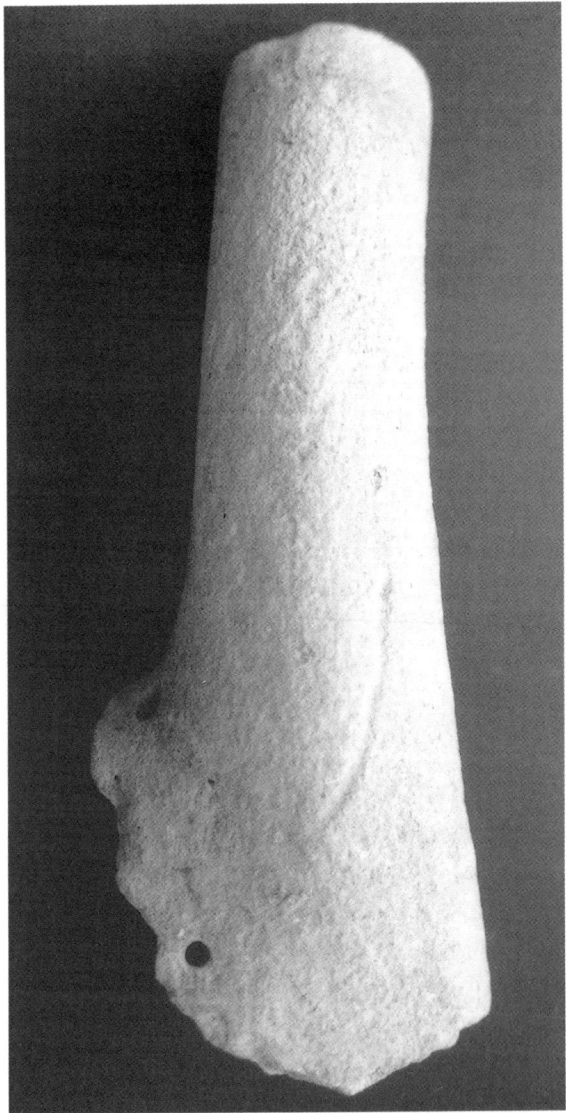

Fig. 6. Handle fragment with hole from repair, Panayia Ematousa, Cyprus.

is the main reason why the shape finally went out of production. Transport amphorae must be easy to handle even when filled up. For this purpose the pointed base was developed, serving as a third handle. With a pointed base the transport amphorae could not stand on their own, but had to be leaned against a wall, as can still be seen in the ruins of Pompeii, or placed in a stand. The efficiency of the amphora is also determined by the relationship between the weight of the container versus the volume. The Cypriot amphorae are simply too big and heavy to be handled comfortably by one person. Another important factor was the suitability of the shape for sea transport. Underwater excavations of ship wrecks have shown that amphorae were placed directly against the hull and stacked vertically, shoulder to shoulder (Tchernia 1978, 21-25, fig. 14). The bases of each consecutive layer of amphorae were placed in the holes between the amphorae in the lower level, and sometimes twigs or branches were used as padding against the hull. The wider the bodies, the more space for the bases, and the slimmer the amphorae, the slimmer the bases had to be. The high, horizontal handles would not have facilitated the stacking.

Though numerous new types of amphorae appeared, the spirit characteristic of the Hellenistic period went towards rationalising. Not only Cyprus, but many other producing areas chose to imitate the most popular amphora types like the Rhodian and Koan ones (Empereur & Hesnard 1987, 13). The Cypriot amphorae with horizontal handles did not live up to the practical standards of the fast-growing economy of the Hellenistic world, and the production ceased.

BIBLIOGRAPHY

Alpözen, T. O. *et al.* 1995
Commercial Amphorae of the Bodrum Museum of Underwater Archaeology. Maritime trade of the Mediterranean in ancient times. Bodrum.

Amyx, D.A. 1958
The Attic Stelai, Part III, *Hesperia* 27, 163-310.

Ben-Tor, A. et al. 1986
Tell Qiri. A Village in the Jazreel Valley. Report of the Archaeological Excavations 1975-1977 (Qedem Monographs 24) Jerusalem.

Brun, J.P. 1933
Discrimination entre les installations oléicoles et vinicoles. Discussions présentées par J.P. Brun, in: M.C. Amouretti & J.P. Brun, (eds.), *La production du vin et de l'huile en Méditerranée* (BCH Suppl. 26) 511-537.

Buhl, M.-L. 1983
Sukas VII, The Near Eastern Pottery and Objects of Other Materials from the Upper Strata. Copenhagen.

Calvet, Y. 1972
Salamine de Chypre III, *Les timbres amphoriques (1965-1970)*. Paris.

Calvet, Y. 1986
Les amphores chypriotes et leur diffusion en Méditerranée Orientale, in: J.Y. Empereur & Y. Garlan (eds.) *Recherches sur les amphores grecques, Actes du colloque international organisé par le Centre National de la Recherche Scientifique, l' Université de Renne II et l'École française d'Athénes*, 1984, (*BCH Suppl.* 13) 505-514.

Courtois, L. 1980
Études pétrographiques, in: J. Briend & J.B. Humbert, (eds.), *Tell Keisan*, Paris, 353-360.

Empereur J.Y. & Hesnard A. 1987
Les amphores hellénistiques, in: P. Lévêque & J.P. Morel (eds.), *Céramiques hellénistiques et romaines* II (*Annales litt. de l'Univ. de Besançon* 331) Paris, 9-71.

Frey, D.F. *et al.* 1978
Deepwater Archaeology. The Capistello Wreck Excavation, Lipari, Aeolian Islands. *IJNA* 7, 279-300.

Gjerstad, E. *et al.* 1935
The Swedish Cyprus Expedition II, *Finds and Results of the Excavations in Cyprus 1927-31*. Stockholm.

Gjerstad, E. 1948
The Swedish Cyprus Expedition IV.2, *The Cypro-Geometric, Cypro-Archaic and Cypro-Classical Periods*. Stockholm.

Gjerstad, E. 1960
Pottery Types: Cypro-Geometric to Cypro-Classical, *OpAth* III, 105-122.

Grace, V. 1979a
Amphoras and the Ancient Wine Trade (Excavations of the Athenian Agora, Picture Book 6), Princeton.

Grace, V. 1979b
Kouriaka, in: V. Karageorghis *et al.*, *Studies Presented in Memory of P. Dikaios*, Nicosia, 178-188.

Gunneweg, J. & Perlmann, L. 1991
The origin of "loop-handle jars" from Tell Keisan, *RBi*, 591-599.

Hadjisavvas, S. 1992
Olive Oil Processing in Cyprus from Bronze Age to the Byzantine Period (Studies in Mediterranean Archaeology 99) Jonsered.

Helmsley, L. MacLaurin 1991
Techniques of the Village Pottery Production, in: J.A. Barlow *et al.*, *Cypriot Ceramics: Reading the Prehistoric Record*, 215-219.

Humbert, J. B. 1991
Essai de classification des amphores dites "à anses de panier", *RBi*, 574-590.

Isager, S. & Skydsgaard, J.E. 1992
Ancient Greek Agriculture. An Introduction, London – New York.

Jacobsen, K. Winther 1998
Transport Amphorae found in 1993-1996, in: L Wriedt Sørensen *et al.*, Third Preliminary Report of the Danish Archaeological Excavations at Panayia Ematousa, Aradippou, Cyprus, *Proceedings of the Danish Institute at Athens* 2, 349-359.

Jacopi, G. 1929
Clara Rhodos III, *Scavi nelle necropoli di Jalisso*, 1924-28. Bergamo.

Jacopi, G. 1931
Clara Rhodos IV, *Scavi nelle necropoli Camiresi*, 1929-30. Bergamo.

Johnston, A. W. & Jones, R.E. 1978
The "SOS" Amphora, *BSA* 73, 103-141.

Karageorghis, V. 1967
Salamis 3, *Excavations in the Necropolis of Salamis* I. Nicosia.

Karageorghis, V. 1970
Salamis 4, *Excavations in the Necropolis of Salamis* II. Nicosia.

Karageorghis, V. 1973
Salamis 5, *Excavations in the Necropolis of Salamis* III. Nicosia.

Karageorghis, V. 1993
The History of Wine in Cyprus. A Brief Survey, in J. Vickers (ed.), *Vines and Wines of Cyprus. 4000 Years of Tradition*, Limassol, 30-45.

London, G. 1990
Töpferei auf Zypern damals – heute. Traditional Pottery in Cyprus. Mainz am Rhein.

Lund, J. 1999
Trade patterns in the Levant from ca. 100 BC to AD 200 – as reflected by the distribution of ceramic fine wares in Cyprus, *MünstBeitr* 18, 1-22.

Marchetti, P. 1978
Terrasse Nord, in: P. Auperts, Rapport sur les travaux de la Mission de l'École Française à Amathonte, *BCH* 102, 946-950.

Markoe, G. 1985
Phoenician Bronze and Silver Bowls from Cyprus and the Mediterranean. Berkeley-Los Angeles-London.

Masson, O. 1967
Les inscriptions syllabiques et alphabetiques de Cellarka, in: V. Karageorghis, *Salamis* 3, *Excavations in the Necropolis of Salamis* I, Nicosia, 269-278.

Niemeier, W. D. 1999
"Die Zierde Ioniens". Ein archaischer Brunnen, der jüngere Athenatempel und Milet vor der Perserzerstörung, *AA*, 373-413.

Parker, A.J. 1992
Ancient Shipwrecks of the Mediterranean & the Roman Provinces (BAR Int.Se. 580). Oxford.

Puech, E. 1980
Inscriptions, incisions et poids, in: J. Briend & J.B. Humbert, (eds.), *Tell Keisan*. Paris, 301-310.

Sagona, A. 1982
Levantine Storage Jars of the 13th to 4th century B.C., *OpAth* 14:7, 73-110.

Salles, J.F. 1980
Niveau 4 (Fer II C), in: Briend, J. & Humbert, J.B. (eds.), *Tell Keisan*. Paris, 131-156.

Salles, J.F. 1991
Du ble, de l'huile et du vin, in: H. Sancisi-Werdenburg & A. Kuhrt (eds.), *Achaemenid History* VI, Leiden, 207-236.

Stern, E. 1973
Material Culture of the Land of the Bible in the Persian Period 538-332 B.C. Warminster-Jerusalem.

Strøm, I. 1992
Evidence from the Sanctuaries, in: I.G. Kopcke & I. Tokumaru (eds.), *Greece between East and West: 10th-8th Centuries B.C.*, Mainz am Rhein, 46-60.

Sørensen, L. Wriedt 1996
Preliminary Report of the Danish Archaeological Excavations at Panayia Ematousa, Aradippou 1993 and 1994, *RDAC*, 135-157.

Tchernia, A. *et al.* 1978
L'Èpave romaine de la Madrague de Giens, Campagne 1972-1975, (Gallia suppl. XXIV).

Whitbread, I.K. 1995
Greek Transport Amphorae. A Petrographical and Archaeological Study (Fitch Laboratory Occasional Paper 4). Athens.

Kristina Winther Jacobsen
Department of Archaeology
and Ethnology
Vandkunsten 5
DK-1467 Copenhagen K

krisjac@hum.ku.dk

THE ONTOGENESIS OF CYPRIOT SIGILLATA*

JOHN LUND

The aim of this study is to examine the emergence and early history of Cypriot Sigillata. It is based on the author's conviction that change in any archaeological material – *in casu* a characteristic ceramic red-gloss ware of the Eastern Mediterranean in the Late Hellenistic and Roman periods – must be due either to deliberate or unconscious choices made by individuals. If so, a proper understanding of these may consequently help us to decipher some of the underlying historical processes. J.W. Hayes, K.W. Slane and other scholars have clarified many of the issues involved, but the question has not yet been the subject of a separate investigation from the perspective of this particular ware.

The ontogenesis of Cypriot Sigillata from the late 2nd century BC to the early 1st century AD is analysed below on the background of the ceramic tradition in its presumed source area: Western Cyprus. An attempt is made to determine whether we are dealing with a development that was primarily local or one brought about by external factors – or a combination of the two. The results are set in a wider context by a brief account of the early development of the other major Hellenistic and Roman red-gloss wares of the Aegean and the Levant.

The countries of the Eastern Mediterranean have been unevenly explored by archaeologists, and the material at hand is accordingly by no means as complete as one might wish for. Also, the available publications of excavations and surveys differ widely in scope and quality. In this as in other fields of research it is all too easy to confuse facts with *factoids*, i.e. "mere speculations or guesses which have been repeated so often that they are eventually taken for hard facts".[1] The conclusions reached below should accordingly be regarded as preliminary and open to revision when the record becomes more fully known.

Terminology

"It has been generally admitted by all who have written on the subject that very considerable difficulties exist in the choice of an appropriate title for the red-glazed ware which is so frequently found on Roman sites in the western provinces of the Empire". F. Oswald and T.D. Pryce wrote these words in 1920,[2] but they still hold true as far as the Eastern Mediterranean is concerned. In 1957 K.M. Kenyon introduced the name Eastern Sigillata for "the red-glazed ware of the Eastern Mediterranean".[3] Cypriot Sigillata is a member of this "family",[4] which also comprises Eastern Sigillata A, B, C and Sagalassos Red Slip Ware.[5] This terminology is by now ubiquitous, but the "sigillata" tag is not without problems.[6]

The term terra sigillata assumed its modern meaning at different points in time in different countries. In Denmark, for instance, it seems to have been first used by the scholar Ole Worm (1588-1654), who applied it to a number of pots in his collection with stamped impressions or relief decoration. They originated from a wide range of places, and few appear to date from the Roman period.[7] Moreover, Worm's *magnum opus*, *Museum Wormianum*, published in 1655, includes a section on different kinds of earth used for sealing, referred to as terra sigillata (**Fig. 1**).[8] The designation was taken literally in those days, and things were hardly different in the inventory of the Danish *Kunstkammer* from 1737, in which the only "terra sigillata" are an Ottoman bowl and "Five Indian Jars of various Shapes".[9] None of the vessels which entered the Collection of Near Eastern and Classical Antiquities in the Danish National Museum in the 19th century were registered under the heading terra sigillata, but in 1895 the German scholar H. Dragendorff characterized the designation as "einem unantiken, aber weit verbreiteten Namen",[10] and he is presumably one of the continental writers, with whom Oswald and Pryce associated the term in 1920.[11]

Strangely, to this day no general agreement exists as to the definition of terra sigillata. To most scholars the words designate a "fine-quality red ware with a glossy red-slipped surface".[12] But this excludes – if taken literally – specimens with a black, red-and-black or grey surface which were presumably produced in the same workshops as the red-gloss ones and are commonly grouped with these.[13] Kathleen Slane is alone in treating the black and grey Eastern Sigillata specimens separately because she perceives them as "different in range of shapes as well

Fig. 1. Terra sigillata as illustrated in the folio *Museum Wormianum*, published in Amsterdam in 1655.

as in technology of production".[14] But seven of the ten shapes in what she defines as a "Black Slip Predecessor" to Eastern Sigillata A are identical with forms current in the latter.

Because of these ambiguities, the present writer adheres to a modified version of a "definizione arbitraria" of terra sigillata, which H. Comfort suggested in 1968: "1) tutto il vasellame da mensa italico e provinciale ricoperto da una superficie rossiccia o rosso-arancione di recente denominata *Glanztonfilm* ...; 2) alcune ceramiche ellenistiche coperte da Glanztonfilm rossiccia o nera che appaiono così strettamente connesse alla classe precedente da non poter essere omesse".[15] This is a good working tool but fails to account for the existence of non-glossy examples of Eastern Sigillata[16] – a dilemma illustrated by Sagalassos Red Slip Ware, which is "considered [both] as a new type of eastern sigillata and also a new type of later Roman red slip ware".[17] Still, the main problem with "Eastern Sigillata" is that the term tends to imply a dependency of sorts on Rome, since terra sigillata is often equated with Italian-type sigillata or other Western Sigillata wares.[18] Of course, archaeologists specializing in Hellenistic and Roman pottery are well aware that Eastern Sigillata A predates Italian-type sigillata by more than 100 years,[19] so the terminology is no real obstacle to research. All the same, it may be a subconscious stumbling block.

CYPRIOT SIGILLATA

Stand der Forschung
Hayes coined the name "Cypriot Sigillata" in 1967 for a distinctive red-gloss ware that was presumably made in Cyprus[20]. His study of pottery from the excavations of the House of Dionysos in Nea Paphos came to fruition in 1985 and 1991 with the publication of what became the standard treatments of the ware with a typology of 59 shapes.[21] Several scholars have subsequently dealt with various aspects of Cypriot Sigillata, and have added a few "new" forms to the repertoire,[22] but Hayes' typological and chronological framework remains largely intact – apart from a few revisions made in 1995 by W.A. Daszewski based on evidence from the Polish excavations at Marina el-Alamein in Egypt.[23] Two years later, Slane published new chronological evidence from the stratified excavations at Tel Anafa in Israel.[24]

The kilns producing Cypriot Sigillata have eluded researchers, and the question of their location remains open. Hayes originally argued that the ware "is far more likely to come from the Soli region [than elsewhere in Cyprus], though the possibility that it is not from Cyprus at all cannot be altogether ruled out".[25] In 1991 he observed that the finds from Nea Paphos "reinforce the theory of manufacture in Cyprus (or, failing that, in an adjacent mainland region)",[26] and he restated the possibility of a source in the Soli region.[27] Subsequent publications of mainland sites have, however, weakened the case for a production centre outside Cyprus, and the present author has argued elsewhere that the geographical and chronological distribution of Cypriot Sigillata rather suggests that the workshops could have been situated in the Nea Paphos region, where the repertoire of shapes is much more fully represented than in any other part of the island. The – admittedly inconclusive – results of clay analyses seem to point in the same direction.[28] A clue to the whereabouts of the workshops may be the fact that the contemporaneous fineware kilns at Pergamon,[29] Ephesos[30] and Sagalassos[31] are located either at the periphery of these cities or in their vicinity – a time-honoured location for such industrial establishments in the Eastern Mediterranean.[32] It might accordingly be an idea to take a new look at the outskirts and surroundings of Nea Paphos in an attempt to – finally – locate the ceramic workshops of the city.

"Forerunners": the last decade of the 2nd century BC

A sequence of stratified – and in some cases coin-dated – deposits at Nea Paphos throws light on the emergence of Cypriot Sigillata.[33] The earliest relevant evidence comes from the so-called Quarry-Pit in Room AΛ in the House of Dionysos, a "uniform deposit of late 2nd century BC (c. 110-100 BC?)".[34] It contained a minimum of 1875 fragments of fine ware. Eastern Sigillata A (1553 specimens) constituted an overwhelming majority, followed by local colour-coated wares,[35] lagynoi and mould-made bowls. For present purposes interest centres on what Hayes refers to as "forerunners" of Cypriot Sigillata:[36] mostly deep bowls with incurved rims (36 specimens, **Fig. 2.1-3**).[37] Other shapes were scarce: the fish-plate (at least three examples) (**Fig. 2.4**), the krater (one example, **Fig. 2.5**), and the lekane (one example, **Fig. 2.6**).[38]

The "forerunners" are rooted in the tradition of Hellenistic ceramics in the Levant – notably that of South Western Cyprus, if Hayes is right in linking the fabric of the incurved-rim bowls to the early series of local Cypriot colour-coated wares.[39] The deep bowl with incurved rim,[40] the fish-plate,[41] and the krater had all but disappeared from the ceramic repertoire in Athens about 150 BC,[42] but these forms remained popular in other parts of the Eastern Mediterranean through the 2nd century BC, if not later. Nea Paphos which was founded about 300 BC became an important centre of pottery manufacture in the Hellenistic period,[43] stimulated presumably by an ever-growing demand for objects of daily use on the part of the many settlers who were attracted to the new city.[44]

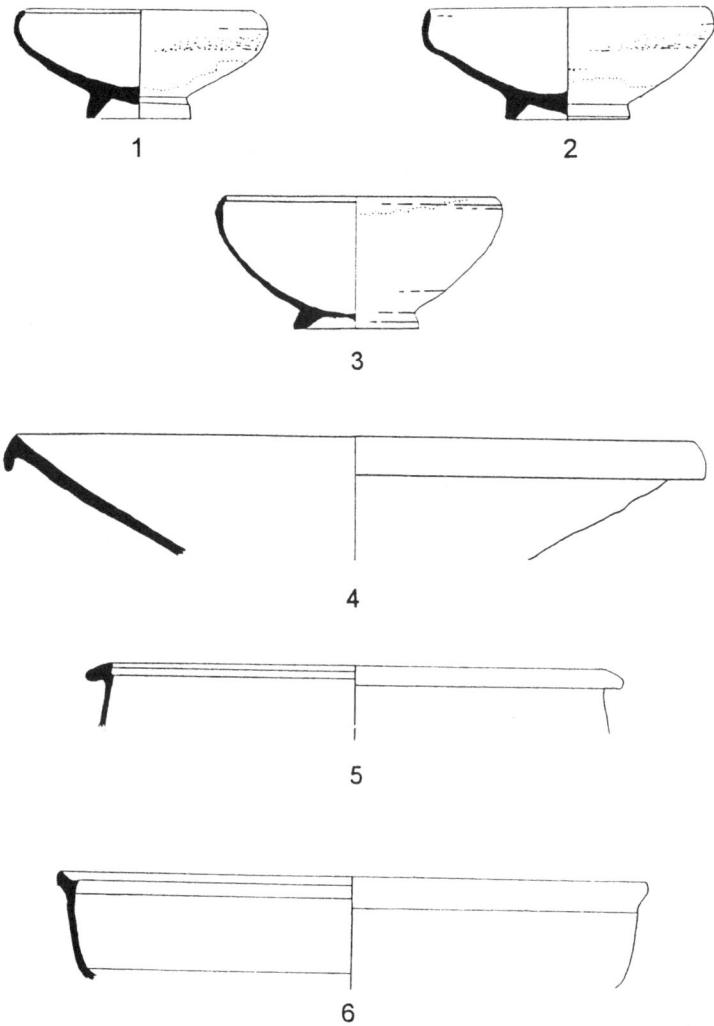

Fig. 2. "Forerunners" of Cypriot Sigillata. Scale 1:3.

The "forerunners" comprise less than 3% of the published finds from the Quarry-Pit. The relatively high incidence of bowls with incurved rim compared with, say, fish-plates suggests that we are dealing with a line of selected forms – not a new complete set of table ware. They could have been targeted at consumers using Eastern Sigillata A who wanted to supplement their red-gloss table ware with the familiar incurved-rim bowl, a form that was – for some reason – not widely marketed in Eastern Sigillata A.

From the late 2nd century to about 31 BC
The first examples of Cypriot Sigillata proper were also found in the Quarry-Pit in the House of Dionysos, a sure sign that the ware was already established by the last decade of the 2nd century BC.[45] It is well documented in contexts through the 1st century BC at Nea Paphos, but the number of finds is not impressive compared with that of Eastern Sigillata A or colour-coated wares in the same stratified deposits.[46] Also, the repertoire of Cypriot Sigillata shapes remained relatively limited and mostly rooted in the local tradition.[47]

The bowls with incurved rim P 20 (**Fig. 3.4**), P 21 (**Fig. 3.5**), the plate P 2 (**Fig. 3.2**), the hemispherical bowl P 18 (**Fig. 4.1-2**), the basin P 36 (**Fig. 4.5**), the lagynos P 44 (**Fig. 5.1**), the amphora of "West Slope Ware" shape P 43 (**Fig. 5.2**), and the slender jug P 49[48] all have parallels among the Middle and Late Hellenistic colour-coated wares of the Eastern Mediterranean – including those manufactured in the local fabric of Nea Paphos[49] – and shapes resembling P 20 and P 36 were indeed represented in the "forerunner" group. The round-bodied strainer jug P 48 (**Fig. 4.6**)[50] is a late descendent of the Hellenistic filter jug, well documented in Athens and elsewhere in the Eastern Mediterranean (including Cyprus) from the middle of the 3rd century BC onwards.[51]

The plate P 1 (**Fig. 3.1**), the mastoid bowl P 15 (**Fig. 3.6**), the small deep bowl with incurved rim P 17 (**Fig. 3.7**) and the hemispherical bowl P 16[52] may have been derived from Eastern Sigillata A models, but they are by no means slavish imitations of those: P 1 is more thin-walled than its supposed prototype, ESA forms 2-4, and lacks its distinctive heavy ring feet. Its base is closer to that of the colour-coated fish plates.[53] P 17, which is quite rare, may be an imitation of ESA 22 A,[54] and P 16, which Hayes compared with ESA 19, is apparently even rarer.[55] Other influences are more difficult to track down: the two-handled cup P 33 (**Fig. 4.4**) may have been inspired by thin-walled cups from the Western Mediterranean such as Mayet 1985 type 2/316,[56] and P 34 (**Fig. 4.3**) could have been derived from P 33. The round-bottomed bowl/krater P 37 (**Fig. 8.9**) has parallels in the red-gloss fabric of Pergamon in the early 1st century BC, but its characteristic astragalos-

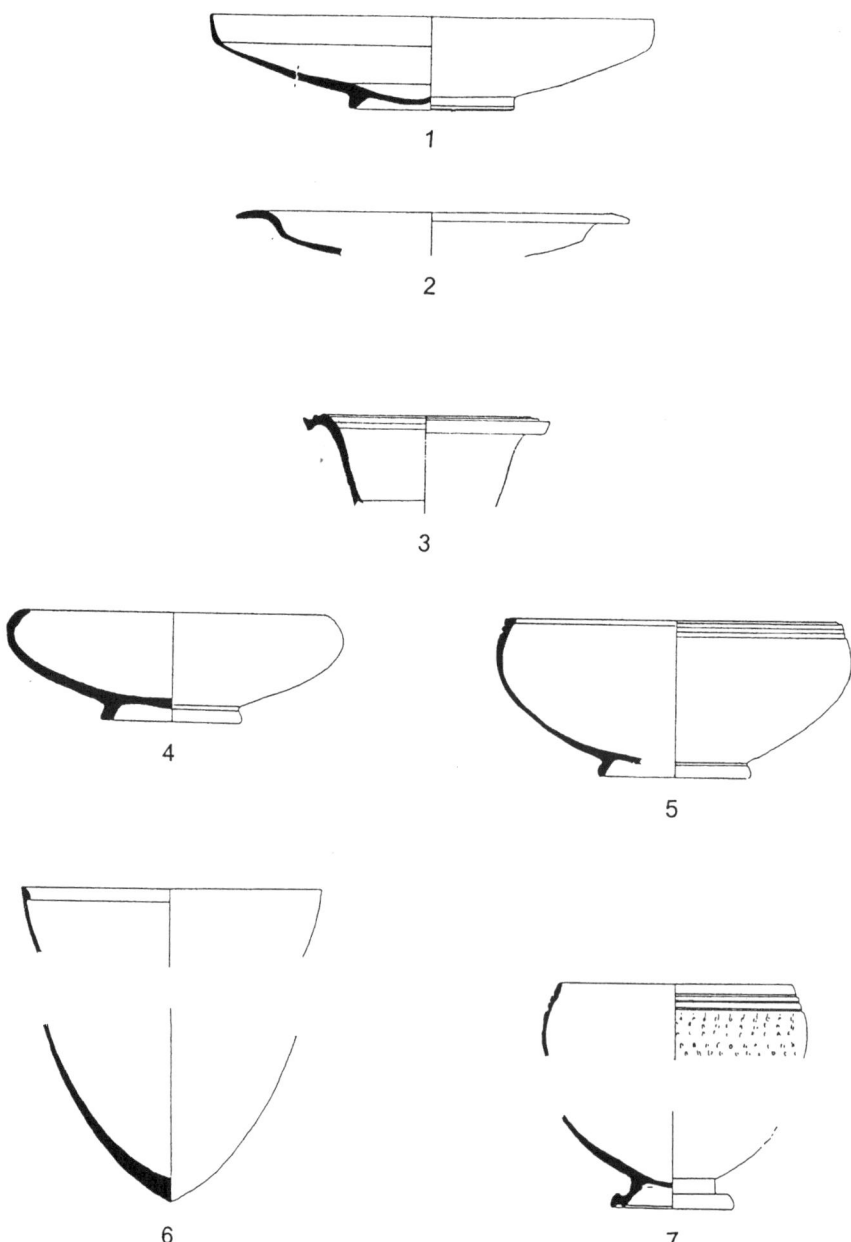

Fig. 3. Cypriot Sigillata. *No. 1*: form P 20; *no. 2*: form P 2; *no. 3*: unclassified form; *no. 4*: form P 20; *no. 5*: form P 21; *no. 6*: form P 15; *no. 7*: form P 17. Scale 1:3.

THE ONTOGENESIS OF CYPRIOT SIGILLATA

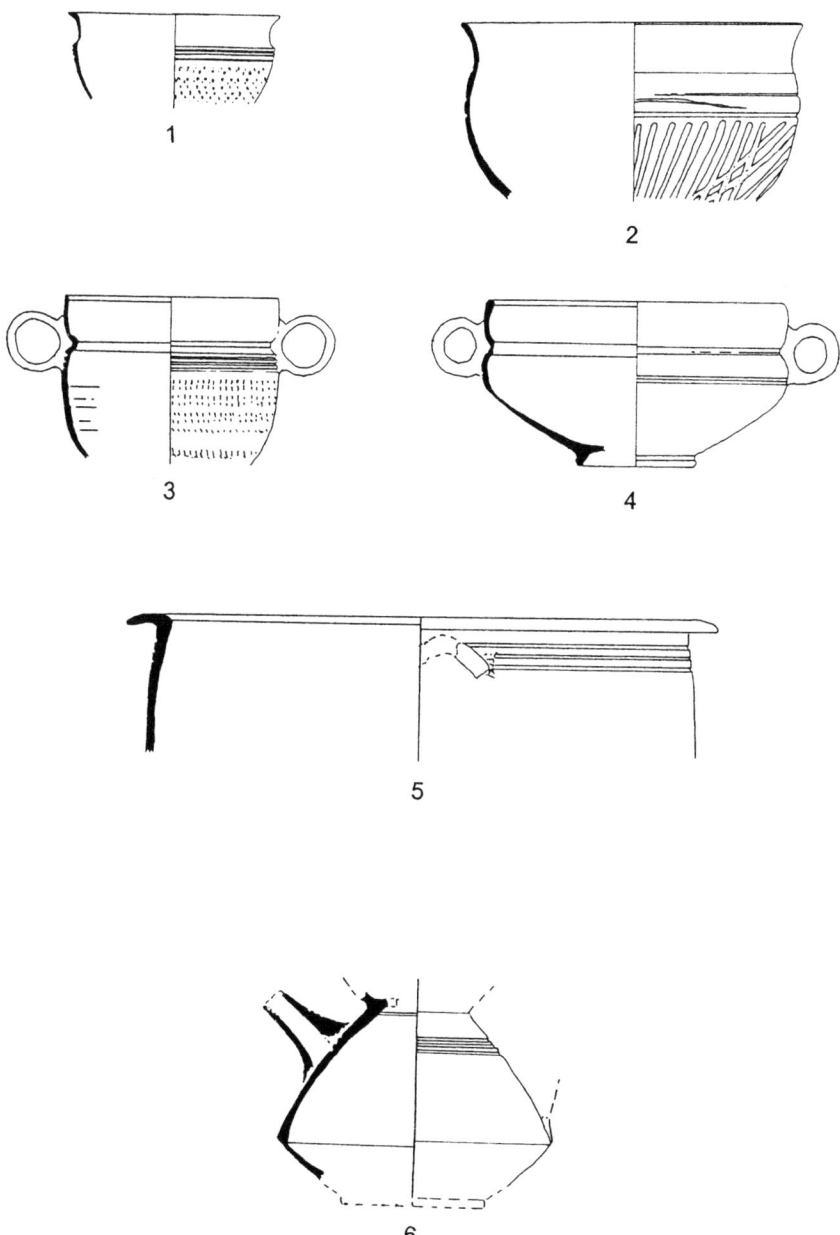

Fig. 4. Cypriot Sigillata. *Nos. 1-2*: form P 18; *no. 3*: form P 34; *no. 4*: form P 33; *no. 5*: form P 36; *no. 6*: form P 48. Scale 1:3.

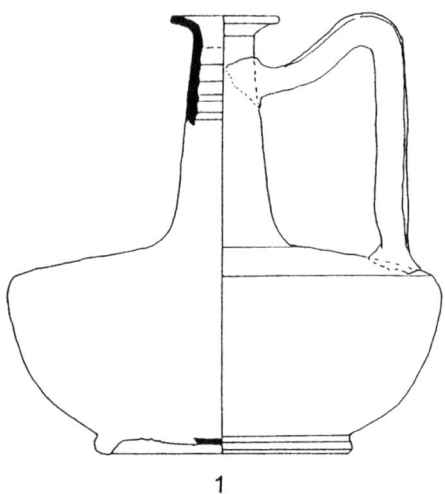

Fig. 5. Cypriot Sigillata. *No. 1*: form P 44; *no. 2*: form P 43. Scale 1:3.

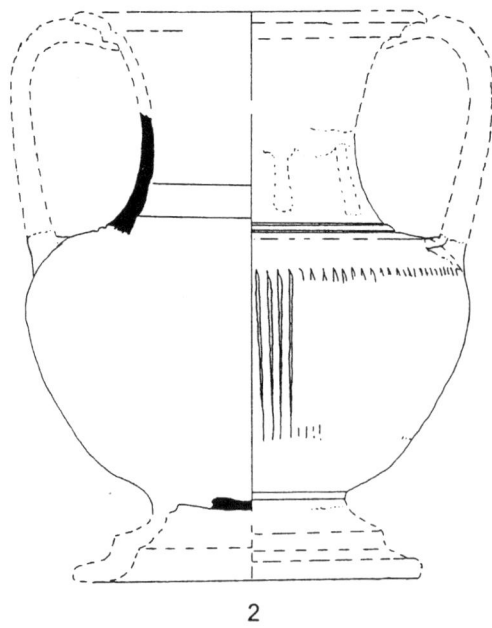

shaped feet are not documented there and seem to be a new and specific Cypriot feature.[57] Hayes regards P 37 A as a large version of the hemispherical bowl P 18 (**Fig. 4.1-2**),[58] but it is not always bigger than its supposed prototype, and it is, perhaps, more likely that the form stems from prototypes such as the chalice represented for instance in the Pergamenian repertoire from the first half of the 1st century BC onwards,[59] unless the prototype should be sought in metal ware.[60]

As before, plates and dishes were apparently not produced in great numbers, but the bowls with incurved rim are at times believed to have been used for serving food – a theory supported by little concrete evidence.[61] It seems more likely that such bowls were (also?) used for wine drinking, which is attested by the lagynos and the amphora of "West Slope Ware" shape.[62] At all events, many of the forms are relatively thin-walled,[63] and it is tempting to see this as an attempt on the part of the potters at creating an alternative to the often thick-walled Eastern Sigillata A ware, which dominated the market in the Levant in the 1st century BC. – P 1, P 17, P 18, P 33, P 34, and the biconical jug with strainer and spout P 47 also occur in grey-fired versions,[64] which Slane regards as a reflection of a "brief transitional (or better, experimental) phase".[65] However, it was not a chronologically distinct entity, for the standard Cypriot Sigillata fabric is – as noted above – already documented in contexts from the end of the 2nd century onwards. Moreover, isolated black gloss examples are not unheard of in the later history of the ware.[66]

The Augustan period I: the emergence of a Cypriot style
The last third of the 1st century BC and the early 1st century AD saw the development of a new range of forms embodying for the first time the imprint of a specific style. It is true that some of the new shapes were apparently only produced in small quantities and quickly dropped out of the repertoire, but others achieved a wide distribution in and outside Cyprus, and the period marks an increase in the exportation of Cypriot Sigillata overseas.[67]

Certain old forms may have been retained for a while: the plates P 1 and 2, the hemispherical bowl P 16 and 18, the two(?)-handled cup P 35, the slender jug P 49, and the bowl/krater P 37 A remained popular until the middle of the 1st century AD. But other forms survived only in a changed form: the hemispherical cup P 19 (**Fig. 8.1**), which is derived from P 18, the round-bottomed bowl/krater X 38 (**Fig. 8.10**), which seems to be a larger version of P 37 (**Fig. 8.9**),[68] and the carinated lagynos X 45 – a successor to P 44.[69]

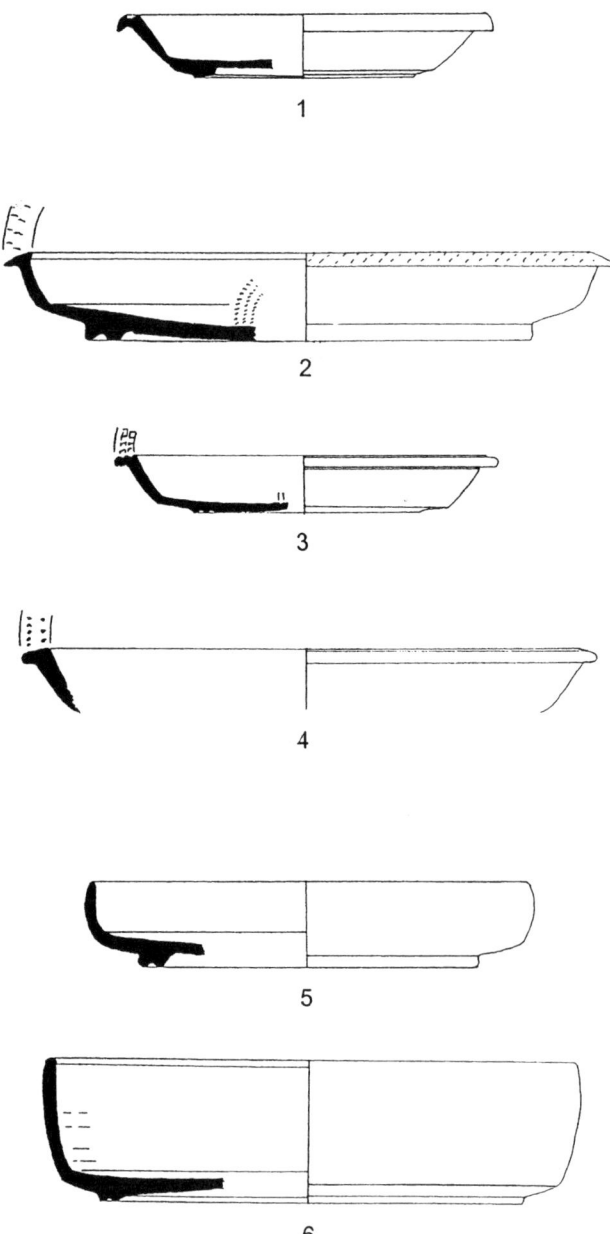

Fig. 6. Cypriot Sigillata. *No. 1*: form P 3; *no. 2*: form P 4 A; *nos. 3-4*: form P 4 A; *no. 5*: form P 5; *no. 6*: form P 6. Scale 1:3.

The introduction of new forms of plates/dishes which were previously rather uncommon, is something of a novelty. The shallow dishes P 3 (**Fig. 6.1**) and P 4 A (**Fig. 6.2-4**) have a distinctive flat rim with rouletting on the top and a double foot ring, and with these go the related round-bodied bowl P 23 (**Fig. 8.3**). Their prototypes may well have been metal vessels: the grooves or small mouldings at the centre of the floor of the slightly later shallow dish P 4 B certainly seem to copy metal ware.[70] They are linked with the rarer dishes P 5 (**Fig. 6.5**) and 6 (**Fig. 6.6**), for which prototypes in Eastern Sigillata A have been suggested, by their distinctive double feet – a feature typical of Cypriot Sigillata.[71] It has been suggested that the bowl with inward-sloping wall and hooked rim, P 22 A (**Fig. 8.8**), may have evolved from P 21.[72] But one might, somewhat speculatively, refer to another possible prototype: a somewhat similar bowl with a hooked and rolled rim (**Plate 12**),[73] which was part of the repertoire of the Red Slip IV (VI) Ware of the Cypro-Classic I period, dated by E. Gjerstad between ca. 475 and 400 BC.[74] A direct connection between the two is out of the question, but we may be dealing with a case analogous to that of certain potters in North Western Asia Minor, who reproduced a 5th century BC. metal ware type in the late 2nd century BC Hayes characterized this as one of "the earliest archaeologically recorded examples of the Roman Republican (or "Late Hellenistic") vogue for imitations of the antique, better represented on later sites such as Pompeii. A model – most likely in bronze or silver – must have been available to inspire the local potters and their presumed patrons. Two local possibilities spring to mind: either an ancient dedication preserved and visible in a local temple ... or else a vase dug out of an ancient tomb in the region ...".[75] It is possible that P 22 had a metal prototype, since its hooked rim seems to be more appropriate for metal than ceramic vessels. The bowls P 27 and P 28 (**Fig. 8.5-6**) which also have a metal feel to them, may likewise be parallelled in Red Slip IV (VI) Ware (**Plate 12**).[76] Is this merely a coincidence?

The Augustan period II: the impact of Italian-type sigillata
An interesting development in the early 1st century AD is the appearance of forms showing clear signs of inspiration – at first or second hand[77] – from Italian-type sigillata. The dishes P 7 (**Fig. 7.1-2**) and 8 (**Fig. 7.3**) were thus loosely based on *Conspectus* forms 12.1 and 18.1,[78] and the

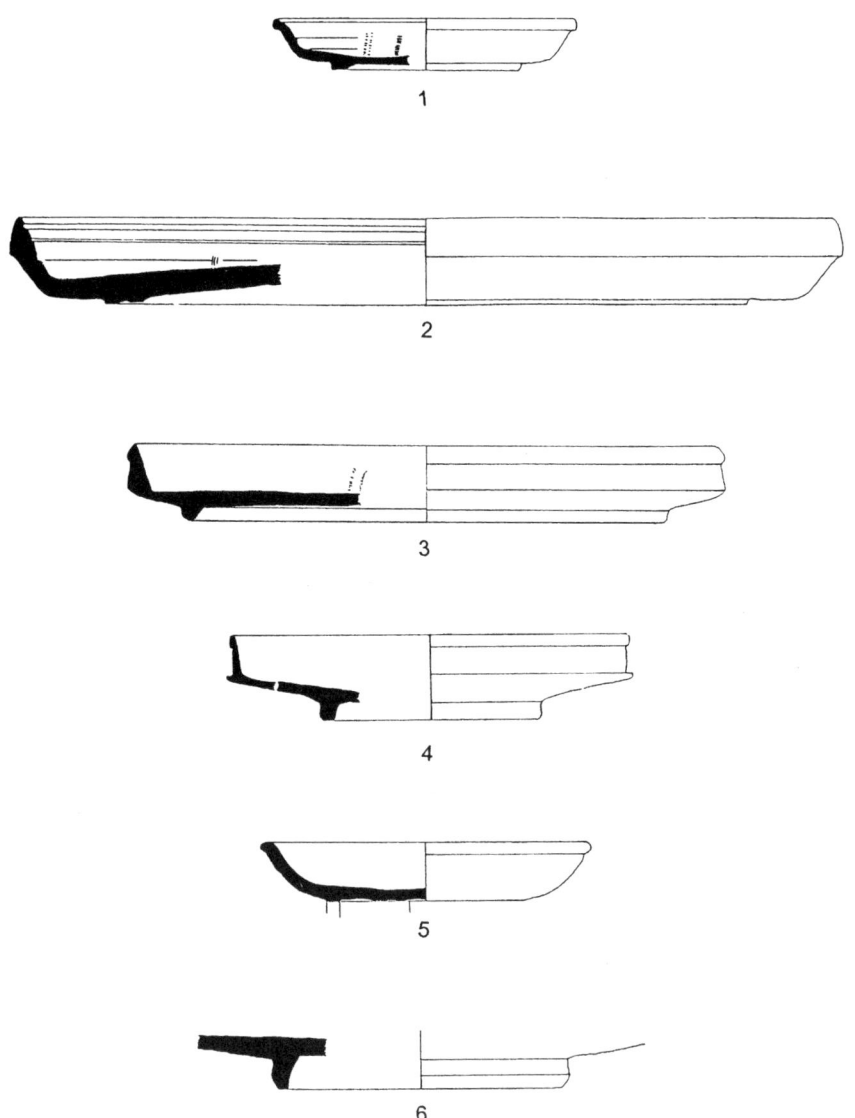

Fig. 7. Cypriot Sigillata. *No 1*: form P 7; *no. 2*: form P 7; *no. 3*: form P 8; *no. 4*: form P 9; *no. 5*: form P 10; *no. 6*: form P 9/13. Scale 1:3.

deep dish P 9/13 (**Fig. 7.6**) on *Conspectus* B 2.5 or the like.[79] The bowl with rounded body P 25 (**Fig. 8.4**) was derived from *Conspectus* form 22,[80] and the same probably goes for the related small hemispherical bowl P 24 (**Fig. 8.2**).[81] The rare footed krater with mouldmade relief decoration X 39 (**Fig. 9.1**) clearly copies the Arretine form Dragendorff 1.[82]

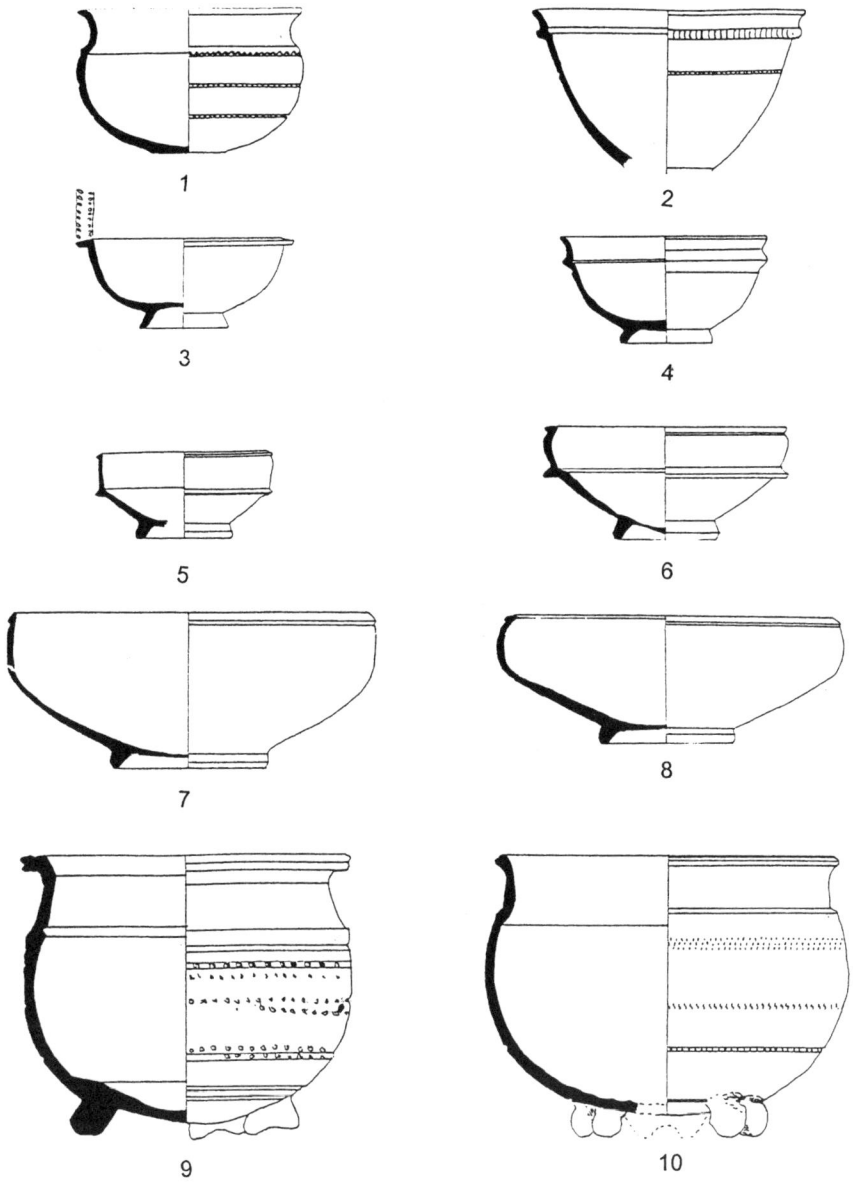

Fig. 8. Cypriot Sigillata. *No. 1*: form P 19; *no. 2*: form P 24; *no. 3*: form P 23B; *no. 4*: form P 25; *no. 5*: form P 27; *no. 6*: form P 28; *no. 7*: form P 21; *no. 8*: form P 22; *no. 9*: form P 37; *no. 10*: form X 38. Scale 1:3.

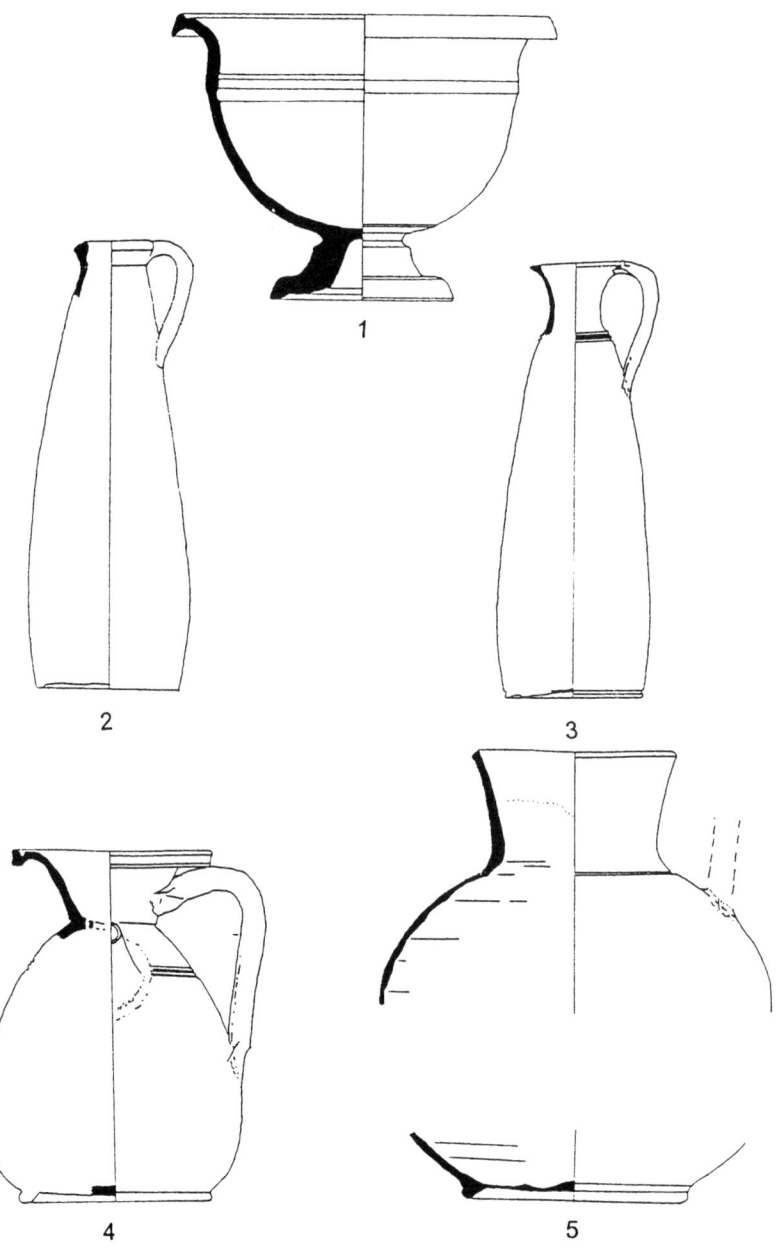

Fig. 9. Cypriot Sigillata. *No. 1*: form X 39; *no. 2*: form X 58; *no. 3*: form P 49; *no. 4*: form P 48; *no. 5*: form P 50. Scale 1:3.

However, we are by no means dealing with exact replicas of the presumed prototypes: the forms have in some cases been altered to bring them in line with other Cypriot Sigillata shapes, as exemplified by the low base of P 7 and the double foot of the large version of P 8. The small dish P 10 (**Fig. 7.5**) may be another case in point, if indeed it is a simplified version of P 7.[83] And none of the forms in question were apparently produced in great quantities.[84]

The occurrence – albeit rare – of potter's stamps is another instance of direct or indirect influence from Italian- type sigillata. The stamps are mostly square or rectangular, but occasionally of the Italian *planta pedis* variety.[85] Nearly all are written in Greek, but the occasional Latin name is also present. One wonders if the short-lived fashion for stamps was sparked off by Italian-type sigillata or by Eastern Sigillata A.[86]

Subsequent developments: a sketch
Production of Cypriot Sigillata increased sharply in the first half of the 1st century AD and reached a peak in the second half of the century.[87] The repertoire of forms was reduced in comparison with the preceding period, but the remaining shapes were turned out in higher quantities than before: the dish P 11 is, for instance, preserved in greater numbers than any other Cypriot Sigillata form. Furthermore the ware was exported as never before (or since).[88]

S.I. Rotroff has asserted that imitation "attests to a fundamental reorientation in the direction of the imitated prototypes and hence bears witness to more meaningful influence [than imports]".[89] If correct, then it is interesting that the forms inspired by Italian type sigillata were hardly produced after the second quarter of the 1st century AD, except for the presumably post-Augustan dish P 9 (**Fig. 7.4**), which seems to have been based on *Conspectus* form 20.4.[90]

May we regard the rejection of the Italian prototypes as a symptom of a conscious or unconscious anti-Roman sentiment? Hardly. More evidence is needed before such a wide ranging conclusion can be drawn. For the time being it must suffice to say that the impact and influence of Italian-type sigillata in Cyprus was short-lived and no more than skin deep.

THE WIDER PICTURE: OTHER EASTERN SIGILLATA WARES

Eastern Sigillata A
The emergence of the first Eastern Sigillata – A[91] – about or shortly before 150 BC was the result of a technological breakthrough.[92] The geographical source of the ware has not yet been identified, but it is increasingly likely that the kilns were situated in North Western Syria or Eastern Cilicia, i.e. in the wider region of Antioch.[93] In an initial phase, black and red-gloss Eastern Sigillata A forms were produced simultaneously,[94] and most of the earliest forms evolved from the mainstream Hellenistic ceramic tradition of the Eastern Mediterranean.[95] The fish plate ESA 1, two plate forms ESA 2 and 6, the mastos ESA 16-17, the bowl with incurved rim ESA 20, and the Megarian bowl ESA 24-25 – are thus red (and occasionally black) gloss versions of preexisting shapes in the Hellenistic repertoire. However these forms were only produced in limited quantities, and by the late 2nd century BC the ware had already found its own unmistakable style, epitomised by the plate form 4 and the footed hemispherical bowl ESA 22,[96] both of which were mass-produced to a hitherto unheard of degree.

J.P. Morel suggested that the plate ESA 6, which is attested in a context dated to 160-140 BC,[97] might copy a prototype from Italy in Campana A ware.[98] This was also proposed in the case of a similar form which is part of the Athenian ceramic repertoire between 150 and 86 BC.[99] However, Slane effectively challenged this theory.[100] To her arguments may be added that Campana A was rarely if ever exported to Northwestern Syria/Eastern Cilicia, and it is difficult to see how potters in that part of the ancient world could have known about this particular form – and whence their incentive to copy it could have come from. Slane plausibly suggests that ESA 6 was "copied from metal",[101] and K. Roth-Rubi already noted in 1984[102] that other forms – the plates ESA 9, 10 and 11 (dated ca. 50-25 BC and 50-20/1 BC) – are copies of metal prototypes. This also applies to an early example of the cup ESA 43 which pre-dates the Late Augustan period.[103] But the examples listed by Roth-Rubi are otherwise relatively late: the cup ESA 45 is dated between AD 1/10 and 50/60, the cup ESA 50 between AD 60/70-100 and perhaps later, and the plates ESA 36-37 between AD 60 and 100.

Eastern Sigillata A proved a huge success on the export markets especially in the 1st century BC and is found – often in great quantities – at most sites in the Eastern Mediterranean. "A new range of vessel forms, derived either directly or indirectly from Arretine ware, gradually ousted the standard Late Hellenistic repertoire" in the first half of the 1st century AD.[104]

Eastern Sigillata C
Eastern Sigillata C[105] grew out of a preexisting tradition for pottery production in Pergamon, centred at first on the potters' quarter of the city, in the valley of Ketios.[106] The bulk of the output did not become red-slipped until about 50 BC[107] and the production moved to Pitane, present-day Çandarli,[108] one of the city's harbours, from the Augustan period onward.

C. Meyer-Schlichtmann, who published the relevant pottery from Pergamon itself, concluded: "Die pergamenische Sigillata dieser Zeitspanne [i.e. the second half of the 2nd century to about 50 BC] läßt sich ohne Schwierigkeiten an die hellenistische Keramik des 2. Jhs. v. Chr. anschließen und führt deren Gefäßauffassung mit geringen Modifizierungen weiter". According to the same author, in the second half of the 1st century BC "findet in der pergamenischen Sigillata ein grundlegender Wandel statt, der an drei Elementen erkennbar ist. Das 1. Element ist [eine] ... Tendenz zu einer strafferen Formgebung, das 2. Element die hohe Anzahl neuentwickelter Typen, darunter hervorzuheben die Entstehung der arretina-ähnlichen Typen. Ältere Typen werden zwar noch in einiger Anzahl weiterverwendet, aber im Sinne der neuen Tendenz umbildet. Kaum merklich und eigentlich erst richtig in der 3. Phase zu erkennen, endet die Tradition hellenistischer Formgebung ... Das 3. Element dieses Wandels ist der überproportionale Anstieg des Anteils der rotbraun und rot gefiernißten Stücke...". Slane criticized some of Meyer-Schlichtmann's conclusions, especially the notion that "Pergamene forms of the mid 1st century BC are the forerunners of Italian sigillata ...". She considers that "the stratigraphic evidence at Pergamon is insufficient to support the suggestion of priority, and none of the pieces comes from a closely dated context".[109] It seems best, then, to regard the "arretina-ähnlichen Typen" as postdating the similar forms in the Italian repertoire. Meyer-Schlichtmann went on to argue that the "Blütezeit der pergamenischen Sigillata mit dem größten Typenrepertoire und einem hohen Qualitätsstandard auf breiter Basis [fell in the first half of the 1st century AD]. Am Anfang dieser Phase enden zwar eine Vielzahl älterer Typen, es werden aber mehr Typen neu entwickelt, die allerdings in ihrer Mehrzahl keine langen Laufzeiten haben und häufig bereits um die Mitte oder im 3. Viertel des 1. Jhs. n. Chr. ... wieder abbrechen".[110] There are many similarities to Cypriot Sigillata in all of this, especially if the chronology is revised as argued by Slane.

Sagalassos Red Slip Ware
With the discovery in 1987 of a potters' quarter at Sagalassos, yet another red-gloss table ware – with a poorly defined Hellenistic antecedent – was added to the Eastern Sigillata family.[111] Production began in the last quarter of the 1st century BC,[112] and J. Poblome has suggested that the arrival of Roman settlers in Pisidia at the time of Augustus "may well have been the impetus for an investment in the already existing local potters' craft ... The floruit of Sagalassos red slip ware seems to have been from the second half of the first century into the third century AD ...".[113]

The first phase in the history of the ware is poorly documented,[114] but it "seems to have been an immediate success since this early material has been found together with other fine wares in early Roman contexts at Perge ... and possibly also at Tel Anafa".[115] Taken as a whole, the ten earliest shapes to be mass-produced at Sagalassos show similarities to a broad spectrum of wares current in the Eastern Mediterranean in Late Hellenistic and Early Roman times. It is rarely feasible to detect a single source of inspiration at this point. The cup 1 A120 and the bowl 1 B170 seem to be rooted in the traditional Hellenistic repertoire, but the influence could conceivably have come second-hand – for instance via Eastern Sigillata A. Other cases in point are the shallow dish 1 C110 and the "container" 1 F150, which have been compared to forms in Italian-type sigillata.[116] However, Poblome noted that a shape similar to the former occurs in Eastern Sigillata A, B, C and Cypriot Sigillata, and the latter is also parallelled in Eastern Sigillata B and Cypriot Sigillata.[117]

Eastern Sigillata B
The ceramic workshops producing the fine ware known as Eastern Sigillata B were established in Asia Minor – presumably in the area of Tralles – in the late 1st century BC.[118] The ware may also have been produced in – and was probably distributed through – Ephesos, an impor-

tant centre for pottery production in the Hellenistic period, if not earlier.[119] But the earliest Eastern Sigillata B forms are unconnected to those of the Hellenistic tradition in the Ephesos region.

In 1904, R. Zahn observed that the name of Caius Sentius, a well-known potter active in the Augustan period, occurred on a stamped example of Eastern Sigillata B from Priene, and he concluded that "C. Sentius eine Filiale in Kleinasien besaß".[120] Hayes later suggested that the Eastern Sigillata B industry could have been established in Asia Minor as branch workshops of Italian entrepreneurs.[121] Ph. Kenrick recently called this theory into question,[122] but there is no doubt that the Eastern Sigillata B shapes were at first deeply influenced by those of Italian-type sigillata.[123] This is amply documented by a deposit with more than 1000 Eastern Sigillata B sherds from the "Tetragonos Agora" in Ephesos, datable between the Middle Augustan and the Early Tiberian periods. S. Zabehlicky-Scheffenegger notes that the earliest forms from here "lehnen sich, wenn auch mit charakteristischen Unterschieden, eng an die italischen Sigillata an ... Die italischen Vorbilder stammen zum überwiegenden Teil aus augusteischer Zeit. Die spätesten Formen ... zeigen größere Abweichungen von den italischen Standardformen, hier liegt schon eine stärkere eigenständige Weiterentwicklung vor".[124]

CONCLUSIONS

Cypriot Sigillata emerged in Western Cyprus – perhaps in the periphery or immediate vicinity of Nea Paphos – in the final decade of the 2nd century BC. It seems to have evolved from the local colour-coated wares of Western Cyprus, since the two are linked by a group of "forerunners". Moreover, many of the earliest Cypriot Sigillata shapes descend from forms current in the Levant in the 2nd and 1st centuries BC. Eastern Sigillata A presumably exerted some influence as well, and the impetus to start a red-gloss production probably came from this ware, which was exceedingly popular in Cyprus in the late 2nd and 1st centuries BC. Production of Cypriot Sigillata appears to have been at a low – but steadily increasing – level through the 1st century BC, and the existence of grey-fired versions of some of the early forms might be evidence of the potters' struggle with the new technique. The output seems to have comprised relatively few plates compared with other forms, perhaps an indication that Cypriot Sigillata was not launched as a complete set of table ware, but rather as a supplement to Eastern Sigillata A.

In 58 BC Cyprus became a Roman province, but this did not at first have any noticeable effect on the production of Cypriot Sigillata. The Augustan period, however, marked a turning point: in the last third of the 1st century BC, a range of new forms was launched, which for the first time embodied a specific "style". They evolved mostly from forms current in the ceramics of Late Hellenistic Cyprus, but apparently also owed a great deal to metal vessels. In a few cases the potters may even have copied centuries old prototypes in the Red-Slipped (and metal?) repertoire of the Cypro-classical period, and it is tempting to see this as a manifestation of a specific Cypriot character in the face of the Roman domination. A return to the roots, so to speak.[125] Be that as it may, certain Cypriot Sigillata shapes were indeed influenced – directly or indirectly – by Italian-type sigillata in the early 1st century AD, and the rare occurrence of potters' stamps may also be due to Italian prototypes. But this influence had already waned before the middle of the 1st century AD: a conscious – or unconscious – rejection of all things Italian?

Seen in a wider perspective there is a clear logic to the emergence and spread of the Eastern Sigillata wares, which has not been lost on most scholars. The earliest ones grew out of Hellenistic ceramic traditions in the Levant,[126] and knowledge of (and interest in) the new technology clearly spread from the east to the west.[127] From North Western Syria/ Eastern Cilicia – where Eastern Sigillata A is believed to have emerged in the second quarter of the 2nd century BC – to Western Cyprus – where Cypriot Sigillata probably began to be made in the last decade of the 2nd century BC – and onwards to North Western Asia Minor, where Eastern Sigillata C became almost exclusively red-glossed about 50 BC.[128]

It is difficult to credit the notion that Eastern Sigillata evolved "unter der politischen Hegemonie der Römer im östlichen Mittelmeergebiet",[129] since Cilicia was an unstable region from the middle of the 2nd century onwards due to the activities of pirates and Rome's efforts to crush them – which finally succeeded in 67 BC.[130] And Syria and Cyprus did not become Roman provinces until 63 and 58 BC, respectively.

True, the Romans had already inherited Pergamon in 133 BC, but Meyer-Schlichtmann is surely right in concluding that "nur als Zufall kann allerdings die historische Parallelität zwischen der Entstehung der pergamenischen Sigillata [Eastern Sigillata C] und dem Übergang Pergamons an Rom bezeichnet werden".[131] Eastern Sigillata B is a different story. It was from the very beginning deeply influenced by Italian-type sigillata and owed little to the local ceramic traditions of Ephesos. The other sigillata wares also came under the influence of Italian type sigillata at some point in the Augustan period, but to varying degrees – a useful reminder of the inter-regional diversity of the Roman Empire.[132]

NOTES

*. This paper went to press on the 4th of September 1999, and publications which have appeared since then could not be taken into account except for brief references in the relevant notes. Moreover, see Poblome and Brulet forthcoming; Meyza forthcoming.

1. Maier 1985, 32.

2. Oswald and Pryce 1920, 3.

3. Kenyon 1957, 281; for an overview of the terminology through time, cf. Meyer-Schlichtmann 1988, 3-9.

4. Cypriot Sigillata is at times referred to as Eastern Sigillata D, cf. Rosenthal 1978, 18-19; Poblome 1999, 291; Schneider 2000, 533-534.

5. The classic treatment of these wares is Hayes 1985. The objections by Bounegru and Erdemgil 1998, 266-267 are difficult to credit at the present time. Hopefully the full publication of the material from the ceramic workshops in the Ketios valley may provide clarification.

6. Cf. Waagé 1974.

7. Schepelern 1971, 330 nos. 4-11.

8. Schepelern 1971, 234-236.

9. Gundestrup 1991, 2-3 i.n. 799/3 and 109, i.n. 818/349.

10. Dragendorff 1895, 19.

11. Oswald and Pryce 1920, 3; cf. further Hediger 2000.

12. Hayes 1997, 41.

13. Cf. for instance Zabehlicky-Scheffenegger 1995, 222. Italian type sigillata was also at first produced in black versions, cf. *Conspectus*, 3-4.

14. Slane 1991, 150, and 1997, 269 with note 53.

15. Comfort 1968, 3.

16. Cf. Slane 1997, 269: "Eastern sigillata A ... The slip ... is usually dark red with a non-reflective sheen". As for Eastern Sigillata C, cf. Meyer-Schlichtmann 1988, 15-17 and Radt 1999, 285: "die sogenannten Pergamenische Sigillata ... eine Keramik mit mattrotem bis glänzendrotem Firnis".

17. Poblome 1999, 25.

18. Cf. for instance Charleston 1955, 14: "of the Western factories"; Garbsch 1982, 7: "im römischen Imperium"; Zabehlicky-Scheffenegger 1995, 217: "Roman"; Zelle 1997, 9 note 49.

19. Hayes 1997, 41; Élaigne 1999; Poblome *et al.* 2000, 280.

20. Hayes 1967, 74.

21. Hayes 1985, 79-91.

22. Cf. Hayes 1991; Daszewski 1995; Meyza 1995; Lund 1997 and 1999; Poblome 1999, 291 and the literature listed in these contributions. A major study by H. Meyza on Cypriot Sigillata and Cypriot Red Slip Ware is in the offing.

23. Daszewski 1995 *passim*. The revisions in the chronology of Cypriot fine wares in the 3rd and 4th centuries AD suggested in Rowe 1998 do not, apparently, affect Cypriot Sigillata – at least not as far as the forms current in the first centuries BC and AD are concerned.

24. Slane 1997, 375-378.

25. Hayes 1967, 74.

26. Hayes 1991, 37 and *idem* 1997, 54: "Cypriot Sigillata is so named because of its prevalence in Cyprus, though some nearby regions have yielded much of it". Cf. Meyza 1995, 179: "there is no decisive proof of their origin in the form of workshops found on the island".

27. Hayes 1991, 27.

28. Lund 1997, 203; *idem* 1999, 3. Schneider 2000, 533-534 with further references.

29. Cf. Erdemgil and Özenir 1982; Bounegru and Erdemgil 1998; Poblome et al. 2001. Radt 1999, 111-112 gives references to earlier literature.

30. Cf. Outschar 1993, 49: "Die Lage an der Peripherie der Stadt in unmittelbarer Nähe einer Hauptverkehrsstraße und des östlichen Stadttors, machen eine Lokalisierung des Töpferviertels in dieser Region durchaus denkbar".

31. Poblome 1999, 24; Poblome et al. 2001.

32. Cf. Oakley 1992, 196-197 fig. 1. Of course, kilns were in many cases located at other places in- and outside the city walls, cf. Arafat and Morgan 1989, 314-329.

33. Hayes 1991; Papuci-Władyka 1995.

34. Hayes 1991, 131: "A late coin of Ptolemy VIII (129?-117/6 B.C.) is present at depth 2.50 m. ... three more of this series ... were found at depth 3.5 m., below 'pithos A'". For the deposit, cf. Hayes 1991, 131-142. As noted on p. 135, the material is unquantified, and the total number of potsherds is unknown, but it must have been huge, judging by the presence of *c.* 1553 fragments of Eastern Sigillata A ware.

35. As in many other cases, "local" is somewhat ambiguous – in this case the word might equally well refer to the Paphos region, Western Cyprus, or, indeed, Cyprus as a whole. For a discussion of the term, cf. Villard 1992.

36. Hayes 1991, 136-137.

37. Hayes 1991, 136-137 nos. 70-72 fig. 49.70-71 and 14.11.

38. Hayes 1991, 137 nos. 74-76 fig. 47. 74-76.

39. Hayes 1991, 27.

40. Cf. Hayes 1991, 136 nos. 57-69 fig. 49.57-69.

41. Cf. Hayes 1991, 135 nos. 41, 44-45 fig. 48.41 and 44-45.

42. Cf. Rotroff 1997a, 162-163, 146-149 and 135-139.

43. Hayes 1991, 26-31 and *passim*; Papuci-Władyka 1995, 247.

44. See Maier and Karageorghis 1984, 224; Lund 1993, 136; Lund and Sørensen 1996, 145-149. Palaipaphos, the predecessor of Nea Paphos, is also thought to have been a centre for pottery production, cf. Maier and von Wartburg 1985, 118.

45. A rim of a bowl of an unclassified form, Hayes 1991, 137 no. 77 fig. 47.77, also two unpublished plate rims, "a mastoid bowl rim (form P 15) and a few other sherds".

46. There is no way of knowing if the published type specimens from the House of Dionysos are representative, but the material from Room BZ, coin-dated between ca. 100 and 60 BC – but with possible later intrusion – comprises Eastern Sigillata A (14 examples), local colour-coated wares (7 examples), painted wares (4 specimens), unguentaria (4 examples), Cypriot Sigillata (3 examples), plain buff wares (2 specimens) and other imports (1 specimen), cf. Hayes 1991, 143-145. The corresponding evidence from layers 4-12 in Room E dated on coin evidence about 50-40/30 BC is as follows: Eastern Sigillata A (14 examples), local colour-coated wares (7 specimens), Cypriot Sigillata (8 examples), Knidian grey ware (1 specimen), cf. Hayes 1991, 147-148.

47. Some of the forms listed below probably continued to be produced and used in the last third of the 1st c. BC.

48. For the forms cf. Hayes 1985, 81-89; *idem* 1991, 38-47. According to Hayes 1991, 150 no. 12 pl. 6.1 the amphora shown here on **Fig. 5.2** is "Early Cypriot Sigillata"; on p. 46 it is referred to as "transitional early ware".

49. P 20 obviously follows the Late Hellenistic tradition of incurved rim bowls, and P 21 has parallels among the imported colour-coated bowls in a mid 2nd c. BC context at Nea Paphos, Hayes 1991, 109 no. 8 fig. 48.8. According to Hayes 1991, 38, the plate P 2 is an "imitation of a Late Hellenistic form (cf. local colour-coated series) ...". The form of P 15, which is derived from the Greek *mastos*, was widespread in the Levant in the Hellenistic period, cf. Greifenhagen 1977. See Waagé 1948, 12 no. 54 pl. 2 for possible Syrian prototypes. Hayes 1991, 151 no. 23 pl. 24 has suggested that a somewhat similar form with a tiny base was probably made somewhere in Eastern Cyprus in the 2nd c. BC, a theory confirmed by the finds from Aradippou: Lund 1996, 144 and 148 nos. 47-48 fig. 5; *idem* 1998, 333 and 336-337 nos. 21-22. More elaborate "Syro-Palestinian" bowls with beaded rim mouldings are discussed by Weinberg 1988. A form similar to that of P 15 also existed in an Eastern Sigillata A version, Hayes 1985, 21 form 18. – P 18 is clearly Cypriot Sigillata version of the Hellenistic mouldmade bowl, well documented in Cyprus, cf. for instance Neuru 1991, 17 no. 26 fig. 8.26. P 36 seems to be a later version of the krater represented among the group of forerunners, cf. **Fig. 1.5**. P 44 is a later version of the previously mentioned lagynoi in the transitional ware. P 43 is better documented in the transitional ware version. Finally, P 49 seems to derive from "a common Cypriot Hellenistic shape ... in colour-coated ware", Hayes 1991, 48.

50. Hayes 1985, 89 and *idem* 1991, 46.

51. Rotroff 1997a, 180-182 with references to occurrences at other sites.

52. Hayes 1985, 81-83 and *idem* 1991, 38-42.

53. Hayes 1985, 81 and 1991, 38.

54. Hayes 1985, 23-24 and *idem* 1991, 41.

55. Hayes 1991, 41.

56. Ricci 1985, 298; a find in the shipwreck at Antikythera, Weinberg *et al.* 1965, 20 n. 7, shows that such bowls were known in the Eastern Mediterranean shortly before the middle of the 1st c. BC; the wreck is dated by coins to ca. 70-60 BC, cf. Yalouris 1990.

57. Radt 1999, 110 refers to astragals as "Würfel-Knöchel mit glückbringender Bedeutung".

58. Hayes 1985, 87 and *idem* 1991, 43.

59. Meyer-Schlichtmann 1988, 168-169 Kg2 pl. 24 and 33, with a suspected life span beginning "in der 1. Hälfte, spätestens um die Mitte des 1. Jhs. v.Chr. Das Ende wird um die Mitte des 1. Jhs. n. Chr. anzusetzen sein ...". Slane suggests that a somewhat similar example from stratum HELL 2B/C at Tel Anafa may be earlier than the first third of the 1st c. BC, Slane 1997, 256-257 no. FW 495 pls. 30 and 53.

60. Hayes 1997, 59.

61. Rotroff 1997a, 161 regards this form as a vessel used for serving food, but no hard evidence for this theory has been adduced, and the lack of handles can hardly be said to speak in favour of the theory: witness the so-called Megarian bowls; cf. also Peignard 1997, 312-313. Hannestad 1983, 15 suggests that the form could have had a wider use.

62. For the former cf. Zanker 1989, 43-55, and for the latter Papuci-Władyka 1992, 40 describing an example probably from Pergamon with painted decoration on the shoulder "in the shape of a grapevine branch rendered in white with red leaves".

63. Hayes 1991, 37. Cf. also Daszewski 1995, 28 and Slane 1997, 366-367.

64. Hayes 1991, 37. Cf. also Slane 1997, 366-367.

65. Slane 1997, 366.

66. Cf. Hayes 1991, 39: P 3; Lund 1993, 102 no. C-159 fig. 44: P 4B.

67. Lund 1997, 207 figs. 6-7 and 210-211; see further Slane 1997, 375: "Of the 153 pieces identified in the second campaign [at Tel Anafa] ... 35 are in ROM 1B strata, 15 in ROM 1B or C, 14 in ROM 1C, 11 in ROM 1C/later, and 67 in late levels; only 11 were from levels classed as earlier than ROM 1B".

68. Hayes 1985, 84 and 87-88.

69. Hayes 1985, 89 and *idem* 1991, 47.

70. Hayes 1991, 39.

71. Cf. Hayes 1985, 81 and Slane 1997, 376. – Hayes 1991, 41 suggested that the dish P 14 might have had a rim as ESA 5, but it is excluded from the present discussion because the form is not with certainty part of the Cypriot Sigillata family, cf. Hayes 1991, 182 no. 26 fig. 61: "just possibly fine early Çandarli ware".

72. Hayes 1985, 84 and *idem* 1991, 42.

73. The Collection of Near Eastern and Classical Antiquities, i.n. 9733 from Marion, Tomb 31: Karageorghis *et al.* 2001, 47-48 no. 87.

74. Gjerstad 1948, 81-82, 199-202 and 427.

75. Hayes 1995, especially 181-182.

76. The Collection of Near Eastern and Classical Antiquieties, i.n. 9735 from Tomb 31 at Marion: Karageorghis *et al.* 2001, 48 no. 88. Slane 1997, 376, on the other hand, refers to the carinated bowl P 28 as an imitation of western sigillata.

77. It should be kept in mind that the potters may in some cases have been inspired by Italian-influenced prototypes in other wares such as Eastern Sigillata A, B or C, cf. Hayes 1985, 85, although this may be an unnecessary hypothesis in view of the fact that Italian type sigillata reached Cyprus in some quantity.

78. Compare Hayes 1985, 82 and *idem* 1991, 39-40 with *Conspectus*, 72-73 and 82-83.

79. Compare Hayes 1985, 83 and *idem* 1991, 41 fig. 18.a with *Conspectus*, 156-157.

80. Compare Hayes 1985, 85 pl. 20.3 and *idem* 1991, 43 fig. 61.

81. Hayes 1985, 85 and *idem* 1991, 43.

82. Hayes 1985, 88 and *idem* 1991, 47.

83. Hayes 1985, 82 and *idem* 1991,40. Hayes suggests that P 10 may be derived from Eastern Sigillata A form 12.

84. Hayes 1985, 82-85 characterises P 7, P 9 as "rara", P 8 and P 24 as "poco comune", P 25 as "non comune" – P 13 is only preserved in "alcuni frammenti".

85. Cf. Hayes 1997, 57.

86. Cf. Poblome *et al.* 2000, 281 for a warning against linking the practice of stamping with the appearance of "Arretina-ähnliche Typen" in the case of Pergamon.

87. Lund 1997, 210-211 fig. 9.

88. Hayes 1985, 80: "una maggiore esportazione di questa produzione iniziò soltanto nella seconda metà del 1 sec. a.C."; Lund 1997, 207.

89. Rotroff 1997b, 98.

90. Compare Hayes 1985, 82 pl. 18.14 and *idem* 1991, 40 figs. 18.1-2 with *Conspectus*, 86-87.

91. Hayes 1985, 9-48; Gassner 1997, 121-123; Slane 1997, 257-258; Zelle 1997, 10-12, 19 and 32-41; Lund 1999; Meyza and Papuci-Władyka 1999; Poblome 1999, 289; Hayes 2000, 287-288.

92. Slane 1997, 269-271; Élaigne 1999.

93. Hayes 1985, 10: "dalla regione siro-palestinese"; Slane 1989: "a north Syrian source"; Lund 1993, 39-40 note 82: "in one or more of the major cities along the Orontes: Antiocheia, Hama and Apamaea"; Slane *et al.* 1994, 64: "a North Phoenician or Syrian source"; Hayes 1997, 54: "somewhere in the Syrian region ... the region of eastern Cilicia north of Iskenderun is now mooted"; Slane 1997, 272: "a northern Syrian source ...The Phoenician sites of Tyre and Sidon ... remain possibilities"; Zelle 1997, 11: "Tarsos ist ein mögliches Zentrum vermutet worden"; Lund 1999, 2: "the wider region of Antioch"; Poblome 1999, 289; Schneider 2000, 532: "the region between Tarsus and Antioch seems a more likely source than Cyprus".

94. *Conspectus*, 3, 5 and 8; however, note Hayes 2000, 288-289 fig. 7.1.

95. Rather than in Greece as suggested by Slane 1997, 273.

96. This form seems to copy an Athenian prototype, cf. Rotroff 1997a, 164.

97. Hayes 1985, 17-18; for the context at Tel Anafa, cf. Slane 1997, 283-285 and 258.

98. Cf. Morel 1976, 492-494. According to Morel there are numerous and evident analogies between Campana B and Eastern Sigillata A.

99. Rotroff 1997a, 13, 154 and 327-328 nos. 838-846 fig. 57; Rotroff 1997b, 99 fig. 1: the form "may have been inspired by Italian imports".

100. Slane 1997, 273.

101. Slane 1997, 283.

102. Roth-Rubi 1984.

103. Cf. Hayes 1985, 11. Rotroff 1997a, 222 notes parallels with the stamped rims of ESA form 9 in calyx kraters from Athens, of which one was found in a context of the second quarter of the 2nd c. BC. She assumes "an eastern origin for the stamped rims".

104. Hayes 1972, 10. At Tel Anafa, "imported sigillatas and ESA forms that imitate western sigillata first appear in ROM 1 B", which has a terminus post quem of "ca. 5 BCE or even later", cf. Slane 1997, 261-262.

105. Hayes 1985, 71-78; Erdemgil and Özenir 1982; Meyer-Schlichtmann 1989; Özyiğit 1990; Gassner 1997, 135-137; Zelle 1997, 13-14, 20-26 and 64-152; Bounegru and Erdemgil 1998; Poblome 1999, 290-291; Poblome et al. 2000, *passim*; Schneider 2000, 533.

106. Cf. *supra* note 29.

107. Slane 1991, 150. For the presumed relationship between the workshops in the Ketios valley and at Pitane, cf. Slane 1991, 150-151; Zelle 1997, 24-26; Poblome et al. 1998, 62 and Radt 1999, 285-287 with further references.

108. Cf. Loeschcke 1912 and Bounegru 1996.

109. Slane 1991, 151.

110. Meyer-Schlichtmann 1988, 198-202.

111. For a comprehensive treatment of Sagalassos Red Slip Ware, cf. Poblome 1999; cf. further Schneider 2000, 533.

112. Poblome et al. 1998, 54.

113. Poblome 1999, 314-315.

114. Poblome 1999, 314: "more archaeological contexts of this early phase need to be excavated".

115. Poblome 1999, 314-315.

116. Poblome 1999, 291: Sagalasssos type 1 C 110 was compared to *Conspectus* form 18 and Sagalassos type 1 F 150 to *Conspectus* form 37.5.

117. Poblome 1999, 289-291.

118. Hayes 1972, 9-10; Hayes 1985, 49-70; Zabehlicky-Scheffenegger 1995; Gassner 1997, 126-135; Zelle 1997, 12-13; 19-20 and 41-64; Lund 1999, 2-3 note 7-8; Poblome 1999, 290; Poblome et al. 2000, 281; Schneider 2000, 532. Cf. also Lund forthcoming.

119. Outschar 1991, 1993 and 1995.

120. Zahn 1904, 437 no. 153 and 444-445.

121. Hayes 1973, 468 and Zabehlicky-Scheffenegger 1995, 222.

122. In a paper read at a conference in Leuven on the 7th and 8th of May 1999: "Early Italian Sigillata – The chronological framework and trade patterns".

123. Hayes 1972, 10: "Eastern Sigillata B ware comes closest to Arretine"; Hayes 1985, 50; Zabehlicky-Scheffenegger 1995, 222. Cf. now also Poblome *et al.* 2000, 281.

124. Zabehlicky-Scheffenegger 1996, 255; cf. also Hayes 1972, 10.

125. For a recent call for "a more balanced attitude towards the question of material remains and political entities", cf. Kletter 1999.

126. Cf. Roth-Rubi 1984, 181; "Diese östliche Sigillata, im Hellenistischen verwurzelt ...".

127. Cf. Hayes 1972, 8: "the beginnings of the red-gloss tradition may be traced in the East"; Slane 1997, 273: "the source of this new technology should be sought in the East".

128. Sagalassos Red Slip Ware seems at first sight to depart from this, in so far as it emerged in Pisidia in the last quarter of the 1st c. BC. But it has now been ascertained that a local tradition for pottery production already existed here in the Hellenistic period, and it is possible that the red-gloss tradition could go further back in the area than what is currently documented.

129. Zelle 1997, 9 note 49.

130. For a re-appraisal of the rôle of the pirates, cf. Rauh 1997.

131. Meyer-Schlichtmann 1988, 208.

132. Cf. also the contrasting situations in Athens and Corinth, Rotroff 1997b, 112-113.

BIBLIOGRAPHY AND ABBREVIATIONS

Alcock, S.E. 1997
The Problem of Romanization, the Power of Athens, in: Hoff, M.C. and Rotroff, S.I. (eds.), *The Romanization of Athens,* (Oxbow Monograph 94), Oxford, 1-7.

Arafat, K. and Morgan, C. 1989
Pots and Potters in Athens and Corinth: A Review, *OxfJA* 8, 311-346.

Bounegru, O. 1996
Notes sur les ateliers de Çandarli, *ReiCretActa* 33, 105-107.

Bounegru, O. and Erdemgil, S. 1998
Terra-Sigillata-Produktion in den Werkstätten von Pergamon-Ketiostal – Vorläufiger Bericht, *IstMitt* 48, 263-276.

Charleston, R.J. 1955
Roman Pottery. London.

Christensen, A.P. and Johansen, Ch.F. 1971
Hama Fouilles et Recherches de la Fondation Carlsberg 1931-1938 III 2: Les poteries hellénistiques et les terres sigillées orientales. Copenhague

Comfort, H. 1968
Terra sigillata, in: Comfort, H., Del Chiaro, A.A. and Paribeni, E., *Terra sigillata. La ceramica a relievo ellenistica e romana*. Roma, 3-38.

Conspectus
Ettlinger, E., Hedinger, B., Hoffmann, B., Kenrick, Ph.M., Pucci, G., Roth-Rubi, K., Schneider, G., Schnurrbein, S. von, Wells, C.M., Zabehlicky-Scheffenegger, S., *Conspectus formarum terrae sigillatae italico modo confectae*. Bonn 1990.

Daszewski, W.-A. 1995
Cypriot Sigillata in Marina el-Alamein, in: Meyza and Młynarzyk (eds.), 27-39.

Dragendorff, H. 1895
Terra sigillata. Ein Beitrag zur Geschichte der griechischen und römischen Keramik, *BonnJb* 96/97, 18-55.

Èlaigne, S. 1999
Le passage des vernis noirs aux vernis rouges. Contribution à l'étude des céramigues fines de Méditerranée orientale, à la lumière du *corpus* alexandrin, *Topoi* 9, 219-228.

Elam, J.M., Glascock, M.D. and Slane, K.W. 1989
A Re-Examination of the Provenance of Eastern Sigillata A, in: Farquhar, R.M. (ed.), *Proceedings of the 26th International Symposium on Archaeometry, Toronto 1988*, Toronto, 179-183.

Erdemgil, S. and Özenir, S. 1982
Preliminary Report on the Kilns excavated in the Ketios Valley, *RdA* 6, 109.

Garbsch, J. 1982
Terra Sigillata. Ein Weltreich im Spiegel seines Luxusgeschirrs. München.

Gassner, V. 1997
FiE XIII/1/1: Das Südtor der Tetragonos-Agora. Keramik und Kleinfunde. Wien.

Gjerstad, E. 1948
SCE IV 2. The Cypro-Geometric, Cypro-Archaic and Cypro-Classical Periods. Stockholm.

Greifenhagen, A. 1977
Mastoi, in: *Festschrift für F. Brommer*, Mainz am Rhein, 133-137.

Gundestrup, B. 1991
Det kongelige danske Kunstkammer 1737/The Royal Danish Kunstkammer 1737 II. København.

Hayes, J.W. 1967
"Cypriot Sigillata", *RDAC*, 65-77.

Hayes, J.W. 1984
Greek and Italian Black-Gloss Wares and Related Wares in the Royal Ontario Museum: A Catalogue. Toronto.

Hayes, J.W. 1985
Sigillate Orientali, in: *EAA Atlante delle forme ceramiche* II. Roma, 1-96.

Hayes, J.W. 1991
Paphos III: The Hellenistic and Roman Pottery. Nicosia.

Hayes, J.W. 1995
Two Kraters "After the Antique" from the Fimbrian Destruction in Troia, *StTroica* 5, 177-183.

Hayes, J.W. 1997
Handbook of Mediterranean Pottery. London.

Hayes, J.W. 2000
From Rome to Beirut and beyond: Asia Minor and eastern Mediterranean trade connections, *ReiCretActa* 36, 285-297.

Hediger, B. 2000
Studia Ietina VIII: Die frühe Terra Sigillata vom Monte Iato, *Sizilien (Ausgrabungen 1971-1988) und frühkaiserzeitliche Fundkomplexe aus dem Peristylhaus I*. Lausanne

Karageorghis, V., Rasmussen, B.B., Sørensen, L.W., Lund, J. Horsnæs, H. and Nielsen, A.M. 2001
Ancient Cypriot Art in Copenhagen. The collections of the National Museum of Denmark and the Ny Carlsberg Glyptotek. Nicosia.

Kenyon, K.M. 1957
Roman and Later Wares. I. Terra Sigillata, in: Crowfoot, J.W. and G.M. and Kenyon, K.M., *The Objects from Samaria*, London, 281-306.

Kletter, R. 1999
Pots and Politics: Material Remains of Late Iron Age Judah in Relation to its Political Boundaries, *BASOR* 314, 19-54.

Loeschcke, S. 1912
Sigillata-Töpfereien in Tschandarli, *AM* 37, 344-407.

Lund, J. 1996
Fine wares from the Hellenistic through the Late Antique periods, *RDAC*, 144-157.

Lund, J. 1997
The Distribution of Cypriot Sigillata as Evidence of Sea-Trade Involving Cyprus, in: Swiny, S., Hohlfelder, R.L. and Swiny, H.W. (eds.), *Res Maritimae. Cyprus and the Eastern Mediterranean from Prehistory to Late Antiquity*, Atlanta, Georgia, 201-215.

Lund, J. 1998
The ceramic finewares from the Late Classical to the Late Antique period found in 1995 and 1996, in: *ProcDanInstAth* 2, 332-348.

Lund, J. 1999
Trade patterns in the Levant from *ca.* 100 BC to AD 200 – as reflected by the distribution of ceramic fine wares in Cyprus, *MünstBeitr* 18 (H. 1), 1-20.

Lund, J. forthcoming
Eastern Sigillata B: a ceramic fine ware industry in the political and commercial landscape of the Eastern Mediterranean, in: Abadie-Reynal, C. (ed.), *Les céramiques en Anatolie aux époques hellénistiques et romaine: production et échanges*.

Lund, J. and Sørensen, L.W. 1996
The Hinterland of the Kingdom of Paphos in the Persian Period. Internal Developments and External Relations, *Transeuphratène* 12, 139-162.

Maier, F.G. 1985
Factoids in Ancient History: the Case of Fifth-Century Cyprus, *JHS* 105, 32-39.

Maier, F.G. and Karageorghis, V. 1984
Paphos, History and Archaeology. Nicosia.

Maier, F.G. and von Wartburg, M.-L. 1985
Excavations at Kouklia (Palaepaphos), Thirteenth preliminary Report: seasons 1983 and 1984, *RDAC*, 100-121.

Meyer-Schlichtmann, C. 1988
Die pergamenische Sigillata aus der Stadtgrabung von Pergamon. Mitte 2. Jh. v. Chr. – Mitte 2. Jh. n. Chr. (PF 6). Berlin and New York.

Meyza, H. 1995
Cypriot Sigillata and Cypriot Red Slip Ware: problems of origin and continuity, in: Meyza and Młynarzyk (eds.), 179-183 and 186-196.

Meyza, H. forthcoming
Cypriot Sigillata and its hypothetical predecessors, in: *Céramiques hellénistiques et romaines. Productions et diffusion en Méditerranée orientale (Chypre, Égypte et côte syro-palestinienne). Colloque du jeudi 2 au samedi 4 mars 2000, Lyon.*

Meyza, H. and Młynarzyk, J. (eds.) 1995
Hellenistic and Roman Pottery in the Eastern Mediterranean – Advances in Scientific Studies. Acts of the II Nieborów Pottery Workshop, Nieborów, 18-20 December 1993. Warsaw.

Meyza, H. and Papuci-Władyka, E. 1999
Nea Paphos, Cyprus: Pottery from cistern STR 1/96-97, in: Machowski, W. (ed.), *Centenary of Mediterranean Archaeology 1897-1997. International Symposium Cracow October 1997.* Kraków, 75-92.

Morel, J.-P. 1976
Céramiques d'Italie et céramiques hellénistiques (150-30 av. J.-C.), in: Zanker, P. (ed.), *Hellenismus in Mittelitalien. Kolloquium in Göttingen vom 5. bis 9. Juni 1974* (Abhandlungen der Akademie der Wissenschaften in Göttingen. Phil.Hist.Kl. 3. Folge N. 97), Göttingen, 471-497.

Negev, A. 1974
The Nabatean Potter's Workshop at Oboda (Acta Rei cretariae romanae fautorum Supplementa 1). Bonn.

Negev, A. 1986
The Late Hellenistic and Early Roman Pottery of Nabatean Oboda. Final Report (QEDEM 22), Jerusalem.

Oakley, J.H. 1992
An Athenian Red-figure Workshop from the Time of the Peloponnesian War, in: Blondé, F. and Perreault, J.Y. (eds.), *Les ateliers de potiers dans le monde grec aux époques géométrique, archaïque et classique* (BCH Supplément 23), Athènes/ Paris, 195-203.

Oswald, F. and Pryce, T.D. 1920
An Introduction to the Study of Terra Sigillata Treated from a Chronological Standpoint. London/New York/Bombay/Calcutta/Madras.

Outschar, U. 1991
Exportorientierte Keramikproduktion auch noch im spätantiken Ephesos? *ReiCretActa* 29-30, 317-327.

Outschar, U. 1993
Produkte aus Ephesos in alle Welt? *Österreichisches Archäologisches Institut. Berichte und Materialien* 5, Wien, 47-52.

Outschar, U. 1996
Beobachtungen und Aspekte zur ephesischen Keramik, in: *Hellenistische und kaiserzeitliche Keramik des östlichen Mittelmeergebietes. Kolloquium Frankfurt 24. - 25. April 1995,* Frankfurt a.M., 35-40.

Papuci-Władyka, E. 1992
A Hellenistic "West Slope" Style Amphora in Warsaw, in: *Studies in ancient art and civilization* 5, Kraków, 39-43.

Papuci-Władyka, E. 1995
Nea Pafos. Studia nad ceramiką hellenistyczną z polskich wykopalisk (1965-1991). Kraków.

Peignard, A. 1997
La vaisselle de la Maison des Sceaux, Délos, in: Δ'επιστημονικη συναντηση για την ελληνιστικη κεραμικη: χρονολογικα προβληματα κλειστα συνολα–εργαστηρια, Αθηνα, 308-316.

Poblome, J. 1999
Sagalassos Red Slip Ware: Typology and Chronology (SEMA 2), Leuven.

Poblome, J., Bounegru, O., Degryse, P., Viane, W. et al. 2001
The sigillata manufactories of Pergamon and Sagalassos, JRA 14, 143-165.

Poblome, J. and Brulet, R. forthcoming
Production Mechanisms of Sigillata Manufactories. When East Meets West, in: Briese, M.B. and Vaag, L.E. (eds.), *Trade Relations in the Eastern Mediterranean from the Late Hellenistic Period to Late Antiquity: The Ceramic Evidence. Ph.D. Seminar for young scholars, Sandbjerg Manorhouse, 12-15 February 1998* (Halicarnassian Studies 3), Odense.

Poblome, J., Brulet, R. and Bounegru, O. 2000
The Concept of Sigillata. Regionalism or integration? *ReiCretActa* 36, 279-283.

Poblome, J., Degryse, P., Librecht, I. and Waelkens, M. 1998
Sagalassos red slip ware. The organization of a manufactory, *MünstBeitr* 17 (H. 2), 52-64.

Radt, W. 1999
Pergamon. Geschichte und Bauten einer antiken Metropole. Darmstadt.

Rauh, N.K. 1997
Who Were the Cilician Pirates? in: Swiny, S., Hohlfelder, R.L. and Swiny, H.W. (eds.), *Res Maritimae. Cyprus and the Eastern Mediterranean from Prehistory to Late Antiquity,* Atlanta, Georgia, 263-283.

Ricci, G. 1985
Ceramica a pareti sottili, in: *EAA Atlante delle forme ceramiche* II, Roma, 231-357.

Rosenthal, E. 1978
The Roman and Byzantine Pottery, in: Stern, E. (ed.), *Excavations at Tel Mevorakh (1973-1976)* (QEDEM 9), Jerusalem, 14-19.

Roth-Rubi, K. 1984
Der Hildesheimer Silberschatz und Terra Sigillata – Eine Gegenüberstellung, *AKorrBl* 14, 175-193.

Rotroff, S.I. 1997a
The Athenian Agora XXIX. Hellenistic pottery: Athenian and imported wheelmade table ware and related material. Princeton, New Jersey.

Rotroff, S.I. 1997b
From Greek to Roman in Athenian Ceramics, in: Hoff, M.C. and Rotroff, S.I. (eds.), *The Romanization of Athens* (Oxbow Monograph 94), Oxford, 97-116.

Rowe, A.H. 1998
A Current Late Roman Site in Nea Paphos, Cyprus, *Near Eastern Archaeology* 61, 179.

Schepelern, H.D. 1971
Museum Wormianum. Dets forudsætninger og tilblivelse. Odense.

Schneider, G. 2000
Chemical and mineralogical studies of Late Hellenistic to Byzantine pottery production in the Eastern Mediterranean, *ReiCretActa* 36, 525-536.

Slane, K.W. 1991
Review of Meyer-Schlichtmann, *Gnomon* 63, 150-154.

Slane, K.W. 1997
The Fine Wares, in: Herbert, S.C. (ed.), *Tel Anafa* II,1 (JRA SupplSer 10 Part II,1), Ann Arbor, Michigan, 247-393.

Slane, K.W., Elam, J.M., Glascock, M.D. and Neff, H. 1994
Compositional Analysis of Eastern Sigillata A and Related Wares from Tel Anafa (Israel), *Journal of Archaeological Science* 21, 51-64.

Villard, F. 1992
Introduction. Les céramiques locales: problèmes généraux, in: Blondé, F. and Perreault, J.Y. (eds.), *Les ateliers de potiers dans le monde grec aux époques géométrique, archaïque et classique* (BCH Supplément 23), Athènes/Paris, 3-9.

Waagé, F.O. 1948
Hellenistic and Roman Tableware of North Syria, in: Waagé, F.O. (ed.), *Antioch on-the-Orontes IV.1: Ceramics and Islamic Coins.* Princeton, 1-60.

Waagé, F.O. 1974
Review of Christensen *et al.* 1971, *AJA* 78, 188-189.

Weinberg, G.D., Grace, V.R., Edwards, G.R. et al. 1965
The Antikythera Shipwreck Reconsidered (Transactions of the American Philosophical Society. NS, 55, 3). Philadelphia.

Weinberg, S.S. 1988
A Syro-Palestinian Bowl Type, *Muse* 22, 64-74.

Yalouris, N. 1990
The Shipwreck of Antikythera. New evidence of its date after supplementary investigation, in: Descœudres, J.-P. (ed.), *ΕΥΜΟΥΣΙΑ: Ceramic and Iconographic Studies in Honour of Alexander Cambitoglou*, Sydney, 135-136.

Zabehlicky-Scheffenegger, S. 1995
Subsidiary Factories of Italian Sigillata Potters: The Ephesian Evidence, in: Koester, H. (ed.), *Ephesos Metropolis of Asia: an Interdisciplinary Approach to its Archaeology, Religion, and Culture* (Harvard Theological Studies 41), 217-228.

Zahn, R. 1904
Scherben von Sigillatagefäßen, in: Wiegand, Th. and Schrader, H., *Priene. Ergebnisse der Ausgrabungen und Untersuchungen in den Jahren 1895-1898*, Berlin, 430-449.

Zanker, P. 1989
Die Trunkene Alte. Das Lachen der Verhöhnten. Frankfurt am Main.

Zelle, M. 1997
Die Terra Sigillata aus der Westtor-Nekropole in Assos (Asia Minor Studien 27), Bonn.

Özyiğit, Ô. 1990
Céramiques hellénistiques d'après les fouilles de Pergame/Kestell in: *Β΄ Επιστημονικη Συναντηση για την ελληνστικη κεαμεικη*, Αθηνα, 94-97.

SOURCES OF ILLUSTRATIONS

Fig. 1. Schepelern 1971, 236.

Fig. 2. Nos. 1-3 after Hayes 1991 fig. 49.71-70 and 14.11; no. 4 after *ibid.* fig. 47.74; no. 5 after *ibid.* fig. 47.76; no. 6 after *ibid.* fig. 47.75.

Fig. 3. No. 1 after Hayes 1991 fig. 52.18; no. 2 after *ibid.* fig. 52.16; no. 3 after *ibid.* fig. 47.77; no. 4 after Papuci-Władyka 1995 pl. 18.115; no. 5 after Hayes 1991 fig. 50.15; no. 6 after *ibid.* fig. 18.15; no. 7 after *ibid.* fig. 18.17 and d.

Fig. 4. Nos. 1-2 after Hayes 1991 fig. 18.2-1; no. 3 after *ibid.* fig. 52.30; no. 4 after *ibid.* fig. 52.Λ.10; no. 5 after *ibid.* fig. 19.36; no. 6 after *ibid.* fig. 49 W.17.2.

Fig. 5. No. 1 after Hayes 1991 fig. 52.17; no. 2 after *ibid.* fig. 52.12.

Fig. 6. No. 1 after Hayes 1991 fig. 18.3; no. 2 after *ibid.* fig. 18.4.1; no. 3 after *ibid.* fig. 61.17; no. 4 from the Danish excavations at Aradippou; no. 5 after Hayes 1991 fig. 18.5; no. 6 after *ibid.* fig. 61.18.

Fig. 7. No. 1 after Hayes 1985 pl. 18.12; no. 2 after Hayes 1991 fig. 18.7.1; no. 3 after *ibid.* fig. 18.8.1; no. 4 after *ibid.* fig. 18.9.1; no. 5 after Lund 1998 fig. 36.52; no. 6 after Hayes 1991 fig. 18.9/13.

Fig. 8. No. 1 after Hayes 1985 pl. 19.11; no. 2 after Hayes 1991 fig. 68.24; no. 3 after Hayes 1985 pl. 20.1; no. 4 after Hayes 1991 fig. 19.25; no. 5 after *ibid.* fig. 66.27; no. 6 after *ibid.* fig. 66.31; no. 7 after *ibid.* fig. 52.19; no. 8 after *ibid.* fig. 19.22; no. 9 after Negev 1986, 27 no. 182; no. 10 after Hayes 1985 pl. 21.2.

Fig. 9. Cypriot Sigillata. No. 1 after Hayes 1985 pl. 21.1; no. 2 after *ibid.* pl. 22.4; no. 3 after Hayes 1991 fig. 20.10; no. 4 after *ibid.* fig. 52.Λ.11; no. 5 after *ibid.* fig. 61.28.

Plate 12. Karageorghis *et al.* 2001, 48.

John Lund
Collection of Near Eastern and Classical Antiquities
The National Museum
Frederiksholms Kanal 12
DK-1220 Copenhagen K

john.lund@nat.mus.dk

A PHIALE IN THE J.F. WILLUMSEN MUSEUM COLLECTION – AN ANALYSIS OF A FORGERY

ANNETTE GABRIELSEN SCHMIDT &
KAARE LUND RASMUSSEN

The Danish artist J.F. Willumsen (1863-1958) was a passionate collector. In 1889 he instigated a private collection of old foreign art (his 'Old Collection'), which came to comprise almost 2,000 inventory numbers. It included 100 items of Greek, Roman, Etruscan and Egyptian antiquities of which 77 are preserved today in the J.F. Willumsen Museum Collection. These objects are mostly terracotta statuettes and bronze figurines, but there are also a few vases and potsherds as well as other artefacts.

In 1996 and 1997 Willumsen's antiquities were exhibited at the Museum of Ancient Art in Aarhus and at the J.F. Willumsen Museum in Frederikssund along with selected works of the artist himself. In connection with the exhibitions the objects were studied and identified by assistant professor Pia Guldager Bilde, Institute of Classical Archaeology, University of Aarhus, and a team of students. It was the first time this had been done in a scholarly context. The research revealed, however, that at least a third of them were modern or pastiches. With the exception of eight modern copies he purchased on purpose, Willumsen believed all his acquisitions to be authentic.[1]

Due to the fact that Willumsen often purchased objects related to artistic or technical problems he was working on, the collection as a whole is of great interest to a student of J.F. Willumsen's art and personality. He seems to have used books on ancient art for inspiration in his own work and not for research. For example, in his copy of Edmund Pottier: *Douris et Les Peintres de Vases grecs*, 1919, he underlined the passages that dealt with technique, instruments and motifs. He also remarks about fig. 20, a drawing of a vase-painting showing restorations in contrasting black and white colours, that it is a strange but decorative use of the colours white on black, black on white. In spite of his own conviction that his eye with time had become so practised that he could always spot a fake, his 'Old Collection' was a mixture of artefacts ranging from masterpieces to fakes.

He was, however, well aware of the fact that the collection was not comparable in standard to those in museums of fine arts, not least because of financial limitations. He thus urged the public to 'appreciate it, for what it was' in the catalogue from the exhibition of his 'Old Collection' in 1947 at the Art Academy in Copenhagen (Willumsen 1947).

A black-figured phiale in the collection was in 1996 identified as Etruscan by the present author (AGS), after some doubts as to its authenticity (**Plate 13**; G.S. 1138; *Tanagra* 1996, no. 63). In 1997 I presented it at the Nordic Vase Symposium 'Ceramics in Context' held at the University of Stockholm where it was further discussed and considered highly suspect. Because of these suspicions it was subjected to thermoluminescence dating at the Radiocarbon Laboratory at the National Museum of Denmark, as well as other scientific investigations that cast light on its authenticity. The doubts proved to be grounded: the phiale is modern.

In the inventory of his collection Willumsen wrote that it was a very small Greek bowl of burnt clay with a depiction of 'a fight in which Zeus, Pallas Athena and others participate with a Medusa head in the middle[2]'. This information probably goes back to the art dealer. Willumsen did not register the objects individually in chronological order but many, *en bloc*, and it is therefore not possible to identify the precise year of purchase. The inventory number G.S. 1138 shows that the phiale must have been bought before 1/10 1927, when he arrived at inv. no. G.S. 1304. His first item for the 'Old Collection' was bought in 1889 and the rather high G.S. -number for the phiale indicates that it was probably bought in the 1920's when Willumsen indeed bought most of his antiquities.

Willumsen carefully saved receipts and often made notes on loose scraps of paper or in his calendar recording a specific purchase. Much of this material has been preserved. On a note recording several purchases, Willumsen wrote (in translation: 'Tanagra. Horse with a Child. Greek. *Cyrénaique*, 125 francs', 'Small Bowl, 50 francs', 'Duck, Egyptian, 50 francs' and 'Mexican Limestone, 25 francs' (**Plate 14**). Of the antiquities mentioned on the note, the 'Egyptian Duck' could be identified with G.S. 1209 in the investigation of 1996 (*Tanagra*, 1996, no. 1), as Mesopotamian, or modern, doubtful because of its cubic shape. The 'Tanagra – Horse with a Child' could be identified with G.S. 1136 (*Tanagra,* 1996, no. 36), genuine. The 'Small Bowl' is probably to be identified with our phiale. Unfortunately there is no record of place or date of purchase on the note, but France is a good guess, since he paid with francs. Willumsen travelled a lot in Europe and lived in Paris 1890-1894 and in Nice from 1918 to 1942. He did not collect during his first stay in France, but during his second stay he purchased several items at auctions, flea markets and from art dealers in Nice, Marseilles and Paris. He bought many of his antiquities, especially the more or less genuine terracotta statuettes, from the art dealer E. Geladakis, 117-18 Palais-Royal, Paris, but in the present case it is not possible to reveal the identity of the dealer.[3]

Description
The phiale is 9 cm across and 2.8 cm high. It has been broken into many pieces and reassembled. The clay is of a yellowish-brown colour. There is a restoration below the Gorgo head, including the bottom of the falling figure. On the outside, the vase has been tinted with a red glaze or wash and at the rim there is a band of black glaze. On the inside the vase has been given a white coating upon which a black-figured decoration has been applied.[4]

The quality and colour of the glaze is, however, very uneven, two of the fragments have a reddish-brown colour and the white coating has an uneven yellowish appearance. There are two stains on the shield of the falling figure, clearly visible after the cleaning of the vase in 1998. They are probably splashes of sealing wax, dropped by mistake when assembling of the fragments.

The decoration on the inside of the vase consists of a Gorgoneion drawn on the central *omphalos*,[5] and circling around this a battle between the gods and the giants and Typhon. Incorporation of Typhon among the giants is rare. In the fourth century BC we have the first examples of Typhon being included in the gigantomachy in art.[6] The snake-legged giants were a result of this mixture, which can be traced in literature as early as Pindar (P. 8,15).

If we would consider it to be genuine (as in *Tanagra* 1996, 147), the vase, despite its archaisms, should be dated sometime early in the second quarter of the fifth century BC, judging from its latest stylistic traits in drapery. That would make it the earliest example of the inclusion of Typhon in the gigantomachy in the visual arts. In other respects the battle has correct iconographic details such as Athena spearing a giant, whom we know from other written and visual representations should be identified as Encelados, and Poseidon fighting with the island Nisyros on his left arm.[7]

The style of the vase displays peculiar traits such as the lengthy limbs, mistakes in drapery (for instance, the odd horizontal incisions in the chiton below Athena's waist) and the incised legs under Zeus' himation and Athena's chiton. These, combined with the poor quality of the black glaze, the archaisms and the late use of the black-figured technique, could have been explained by an Etruscan origin[8], but more or less the same reasons also point to a modern origin. Another stylistic trait that strongly points to a modern origin is the stiff, almost statuesque, depiction of the figures and their isolation from each other.

The artist/forger of the phiale must have used ancient depictions of gigantomachies as his inspiration or models. It has not been possible to identify specific models although the three-bodied Typhon might have been inspired by the 'Typhon'- pediment sculpture found on the Athenian Acropolis in 1888 and identified as Typhon in 1889 by A. Brückner. The only other example in Greek art of a three-bodied Typhon is shown *en face* on a black-figured cup in Florence[9].

The artist/forger probably included Typhon in the gigantomachy out of ignorance. Perhaps he believed that the snake-legged monster without wings was a giant and the creature with wings was Typhon to the Greeks, a theory by Heinrich Heydemann in 1876 (13). This would be a *terminus post quem* for the date of the fabrication and might explain why the artist left out the wings, if the Typhon from the Acropolis was indeed his model.

Trace Element Chemical Analysis
The phiale, as it entered the Old Collection, exhibited a rather crude assembling. It had previously been broken into some 20 pieces, two of which have a reddish-brown tint (**Fig. 1**). In 1998 the Department of

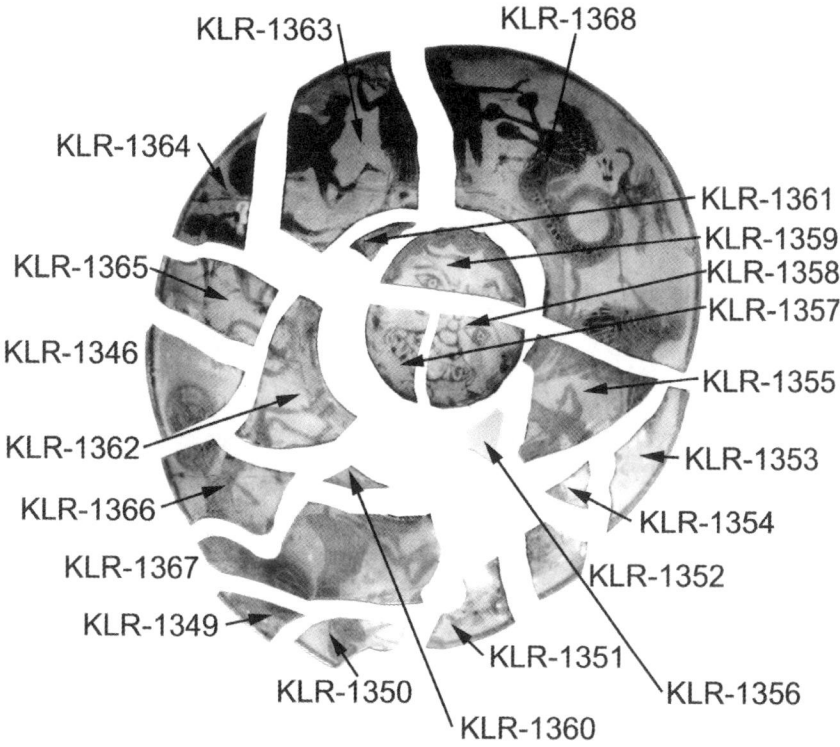

Fig. 1. The phiale before it was taken apart at the National Museum of Denmark. The laboratory numbers of the individual pieces are shown (photo by Benni Berg and Kaare Lund Rasmussen)

Conservation at the National Museum of Denmark disassembled the phiale, cleaned and reassembled it for the exhibition. In the process, a 1 mm hole was drilled with a hand-held high-speed steel drill and a sample of 86.1 mg of ceramic powder was taken from the bottom fragment, KLR-1359, under TL/OSL-protective light.

Instrumental Neutron Activation Analysis (INAA) was performed on a 0.48 mg sub-sample, in order to determine the uranium (U), thorium (Th), and potassium (K) for thermoluminescence dating and for yielding the chemical composition of the clay. The analysis is shown in **Fig. 2**. Neither uranium nor potassium was above detection limits. However, thorium was accurately determined to a concentration of 13 mg/g. **Table 1** also gives analyses of the international standards Marine-

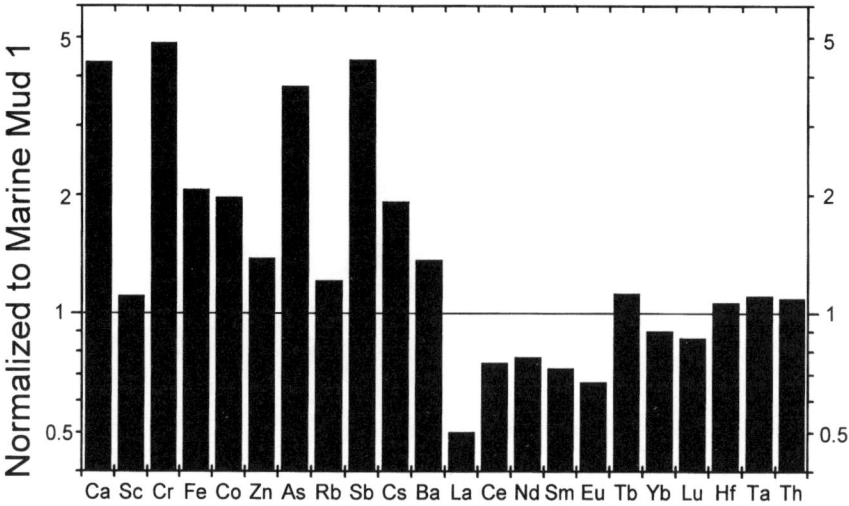

Fig. 2. Abundances of the phiale normalised to the standard sample Marine Mud 1. Note that the y-axis is logarithmic. Gold is left out because it is way out of scale, probably due to a later contamination.

Mud-1 and Cody-Shale-1 (SCO-1), and these show that our sample has a Th-composition resembling Marine-Mud-1. The Th/U-ratio for Marine-Mud-1 is 4.4 and if we assume this ratio to be the same in our sample we obtain an estimated uranium content of 2.9 mg/g. Likewise, if we assume a K/Th-ratio of 1235 similar to Marine-Mud-1, we obtain an estimated potassium content in the phiale of 1.61 wt%.

The composition of the phiale resembles Marine-Mud-1, and this also applies to the other elements analysed. The phiale is thus probably made of pure clay. The concentrations found in the phiale are shown as ratios to Marine-Mud-1 in **Fig. 2**. Four elements, calcium (Ca), cromium (Cr), arsen (As), and antimony (Sb), are significantly higher in the phiale than in Marine-Mud-1 and this might provide a clue to the provenance of the clay. All four elements are enriched by approximately a factor of 4.

Thermoluminescence Dating
The thermoluminescence method is based on the fact that ceramic materials store electrons released by natural radioactivity in traps, and a subsequent laboratory heating causes a release of the stored energy in

the form of light, thermoluminescence (Aitken 1990, 141; McKeever 1985, 253). Kiln firing will set a time-zero in the pottery's energy storage and what is measured in the laboratory is the re-accumulated energy since firing. This provides a dating and it is a well-suited method to expose fakes.

A 5.9 mg sub-sample of the powder (KLR-1348) was subjected to TL-analysis on a TL-DA-12 TL/OSL-system. The sample powder was not subjected to grain-size sorting nor to chemical pre-treatment. A regeneration procedure was run on a single aliquot of the sample (5.9 mg). The dose equivalent was estimated to be 520 ± 40 mGy. As outlined above, the uranium, thorium and potassium contents are estimated to be 2.9 mg/g, 13 mg/g and 1.61 wt% respectively. Assuming that the palaeo-surroundings of the phiale were of a similar composition, that the exterior dose was 150 mGy/y (Kolstrup and Mejdahl 1986), and that the water saturation was 5%, we arrive at an age of 100 ± 30 years BP (before present), or an absolute date of AD 1900 ± 30 (KLR-1348). It should be noted that this date must be considered only a rough estimate of the age, since no chemical pre-treatment was performed, and since the U and K concentrations are estimated values. The uncertainty has accordingly been multiplied by a factor of 2.

Even considering the uncertainties in the dating, an age of the earlier 5th century BC, for example, is possible only if the phiale has been re-heated to annealing temperatures in excess of ca. 400°C within the last century. An event we consider unlikely.

We should, however, issue one *caveat* with respect to the date, namely that we have dated only the central fragment (showing the Gorgoneion). It is possible, if perhaps not very likely, that the central fragment is a later repair, whereas the thin sides of the phiale are genuine. In order to prove or disprove this option we would have to sample one of the side fragments for at least 1 mg of material, which has not been done so far. Having studied the phiale in detail we do not, however, consider this option very likely. To our best judgement the date of AD 1900 ± 30 reflects the age of the entire phiale.

Magnetic susceptibility measurements
In order to test the nature of black glaze magnetic susceptibility was measured on all fragments except the largest (KLR-1368), which was measured on a Kappabridge KLY-2 magnetic susceptibility meter. The

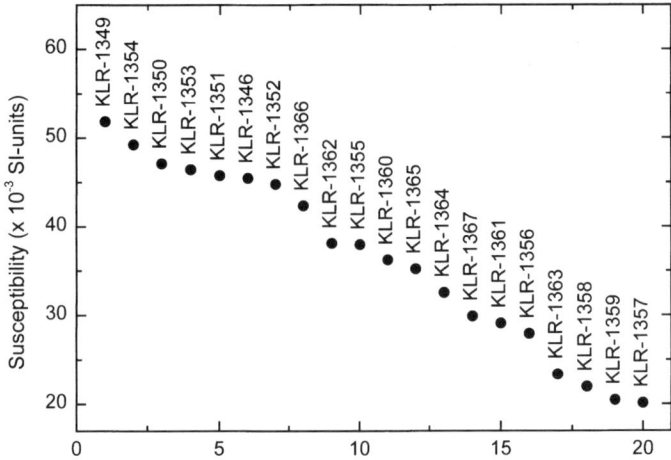

Fig. 3. The magnetic susceptibility (in 10^{-3} SI-units) normalised to a sample volume of 10 cc using the sample weight and an assumed density of 2.5 grams/cc of all the fragments except the largest fragment.

results, normalised to a density of 2.5 gram/cc and a sample volume of 10 cc, are given in **Table 2** and show a rather large variation, between 20 and 53 x 10^{-3} SI-units (**Fig. 3**).

It is interesting to note that the magnetic susceptibility is negatively correlated with the weight of the fragments (**Fig. 4**). An explanation for this correlation could be that the amount and/or grain size of the hematite is correlated with the firing temperature and that the parts that reached the highest firing temperatures could be the most brittle and therefore fragmented into the smallest pieces. The position of the fragments in the phiale augments this interpretation (**Fig. 1**). It seems that the firing temperature has been slightly lower on the right side of **Fig. 1** and highest in the bottom left. The cause could well be a temperature gradient in the firing oven.

If the black glaze was produced using the ancient technique of the Greeks and Etruscans, it should include magnetic minerals, such as magnetite (Noble, 1965, 41). In order to investigate this we have estimated the approximate percentage of black area on the fragments and plotted it *versus* the magnetic susceptibility (**Fig. 5**). This shows a weak negative correlation, so it is highly unlikely that a magnetic mineral such as magnetite was included in the black glaze.

A PHIALE IN THE J.F. WILLUMSEN MUSEUM COLLECTION

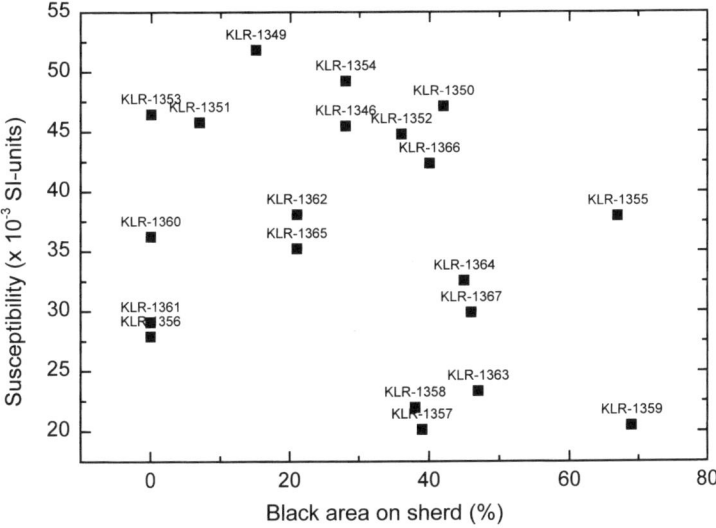

Fig. 4. Magnetic susceptibility as a function of sample weight. A weak negative correlation is seen.

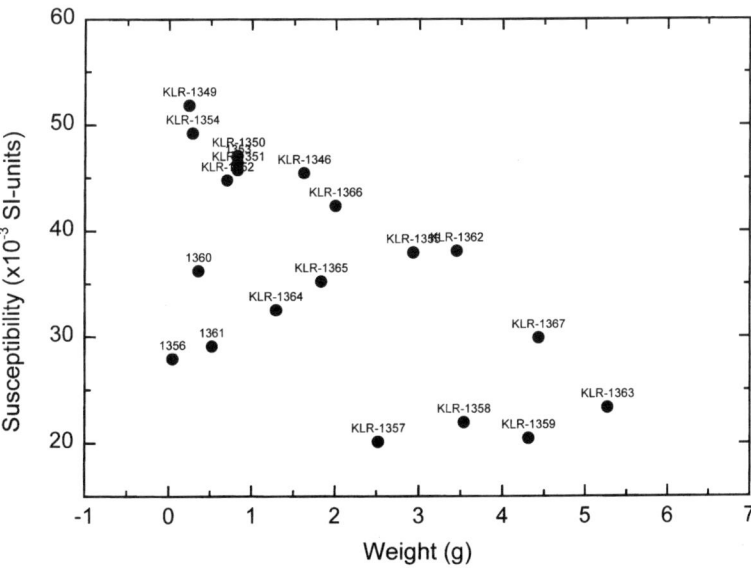

Fig. 5. Magnetic susceptibility as a function of the approximate percentage area covered with black paint on the fragment. A weak negative correlation is seen.

The percentage of black glaze does, however, show a weak positive correlation with the weight of the fragments, as is shown on **Fig. 6**. This could be a coincidence, but it could also somewhat jeopardise the suggestion stated above about a gradient in the firing temperature of the oven. An alternative explanation could be that the correlation is due to a glue-effect of the black glaze.

It is noteworthy that the two fragments tinted reddish brown (KLR-1355 and KLR-1367) do not exhibit an abnormally high susceptibility, indicating that the tinting is probably not due to hematite or magnetite. This indicates that the colouring agent must be a compound with low magnetic susceptibility, excluding several strongly coloured iron-containing compounds otherwise normally found in ground water. This gives rise to a suspicion that the two fragments have been artificially tinted, perhaps by accident or perhaps on purpose, during the time when the phiale was broken.

The composition of the black glaze
In order to elucidate the composition of the black glaze further we removed a few sub-milligram samples of the black glaze with a scalpel. Nine small samples were transferred to a double adhesive tape on a Scanning Electron Microscope sample stub. The samples were analysed by the energy dispersive detector of the SEM (**Fig. 7** and **Table 3**). The quantitative analysis showed small amounts of Mn (0.25 ± 0.06 wt%) and relatively large amounts of Fe (17.9 wt%). The Fe is probably residing in the silicates and in small amounts of hematite that are normally formed during firing. Manganese is, however, very rare in silicates, and is almost certainly an indication of the presence of manganese dioxide (MnO_2 in the black glaze. This fits well with the vanishing correlation between susceptibility and percent black glaze area.

The presence of manganese in the black glaze is very interesting. The ancient technique of making the black glaze was not rediscovered until during the Second World War by Dr. Theodor Schumann. Before that other means had to be employed to imitate it, and one of these was to use MnO_2.

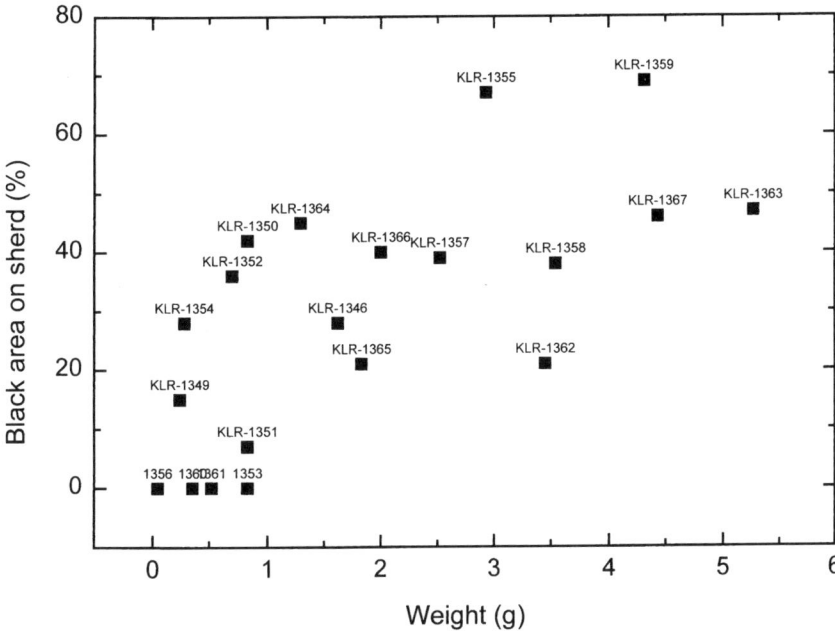

Fig. 6. Percentage of black paint as a function of sample weight. A weak positive correlation is seen.

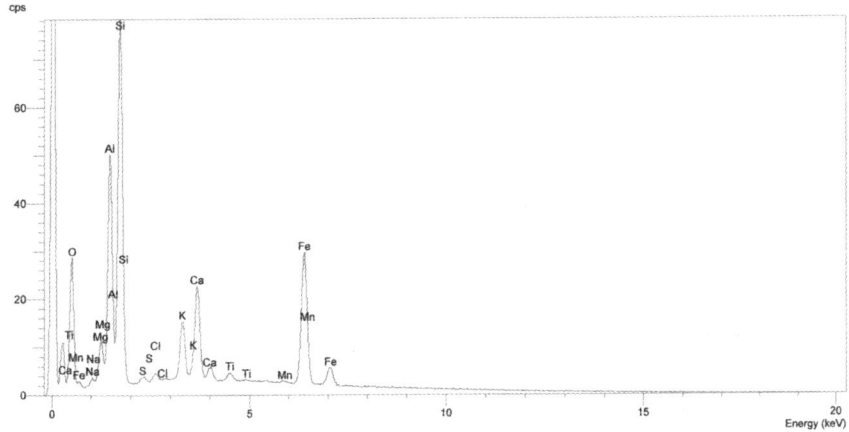

Fig. 7. The energy dispersive X-ray spectrum of nine small samples of black glaze. The acceleration voltage was 20 kV. Pronounced peaks are marked. Note that manganese is also detected.

For example, the workshop of Riccardo Riccardi and his cousins Teodoro and Vigilio Riccardi in Orvieto used MnO_2 for paint and glazes when they were producing forgeries of Etruscan terracotta reliefs and statues during the 1910's, as described by the Alfredo Fioravanti, who worked there (Andrén 1986, 71). Their terracotta statues of Etruscan warriors, acquired by the Metropolitan Museum of Art in 1915, 1916 and 1921, were considered by most to be ancient masterpieces, until definitively exposed as fakes in 1961 (Bothmer & Noble 1961). The Tarquinian imitator of Attic and Etruscan pottery, Scapini, also employed cobalt, lead and manganese in his excellent reproductions of Greek black glaze (Bothmer & Noble 1961, 21).

None of the black glaze figures show any sign of fluorescence when viewed in UV-light. This indicates that the phiale has been burnt thoroughly after the black glaze has been applied. This makes it possible, however unlikely, that the figures have been painted in recent times on an simple, ancient phiale, which was then reheated. To apply a painted decoration on a plain, ancient vase is a common earlier method of falsification (cf. *Fakes* 1973, no. 4), but to reheat it seems excessive.

Conclusion

To sum up we can conclude that the vase was made in what has been called 'the great age of faking', that is, the 19th century and the beginning of the 20th century – up until the 1930's (*Fake?* 1990, 161-162). At that time, collecting antiquities began to spread from the aristocracy to other social classes. Combined with romanticism this resulted in an immense demand for antiques and as a result, many craftsmen and artists turned to forgery. Willumsen was certainly not the only one deceived – today and then, this also happens to scholars and museum curators. Fakes are in fact excellent indicators of the popularity of an item.

Forgery involves the *intention* to deceive and not just imitation or copying. The phiale was clearly sold to Willumsen as a genuine antiquity – not only did Willumsen understand it as such, but the 'restored look' is also telling evidence.

TABLE 1

	Phiale ppm	Marine Mud 1 ppm	Cody Shale 1 ppm	Ratio Phiale/MM-1
K	n.d.	1.47 wt%	1.15 wt%	–
Ca	4.3 wt%	0.98 wt%	1.87 wt%	4.36
Sc	19	17.2	10.8	1.11
Cr	472	97	68	4.87
Fe	4.9 wt%	2.38 wt%	1.80 wt%	2.07
Co	40	20.4	10.5	1.98
Zn	180	130	103	1.38
As	35	9.2	12.4	3.79
Rb	181	149	112	1.21
Sb	4.2	0.96	2.5	4.42
Cs	17	8.6	7.8	1.93
Ba	656	479	570	1.37
La	22	43	29.5	0.50
Ce	66	88	62	0.75
Nd	30	38	26	0.78
Sm	5.5	7.5	5.3	0.73
Eu	1	1.55	1.19	0.67
Tb	1.1	0.96	0.7	1.13
Yb	2.4	2.6	2.27	0.90
Lu	0.3	0.4	0.34	0.87
Hf	4	3.7	4.6	1.07
Ta	1.2	1.1	0.92	1.11
Au	0.099	0.0024	0.0021	41.2
Th	13	11.9	9.7	1.09
U	n.d.	2.7	3.0	–

Results of Instrumental Neutron Activation Analysis of sample KLR-1348 given in ppm (µg/g) where nothing else is indicated. n.d.: not detected. The sample weight was 0.48 mg. The irradiation was done at the light water reactor ASTRA in Zibersdorf, Austria, in a neutron flux of $8 \cdot 10^{13}$ cm^{-2} sec^{-1}.

TABLE 2

	Weight (g)	Susceptibility x 10^{-3} SI-units	Approximate area of black paint (%)
KLR-1346	1.62	45.5	28
KLR-1349	0.2394	51.9	15
KLR-1350	0.83	47.1	42
KLR-1351	0.83	45.8	7
KLR-1352	0.70	44.8	36
KLR-1353	0.83	46.5	0
KLR-1354	0.28	49.2	28
KLR-1355	2.93	38.0	67
KLR-1356	0.0452	27.9	0
KLR-1357	2.52	20.1	39
KLR-1358	3.54	22.0	38
KLR-1359	4.31	20.5	69
KLR-1360	0.3534	36.3	0
KLR-1361	0.52	29.1	0
KLR-1362	3.45	38.1	21
KLR-1363	5.27	23.3	47
KLR-1364	1.29	32.6	45
KLR-1365	1.83	35.3	21
KLR-1366	2.00	42.4	40
KLR-1367	4.43	29.9	46

Table 2. Magnetic susceptibility normalised to a sample density of 2.5 grams/cc and a 10 cc sample volume. The area covered with black paint is roughly estimated by visual inspection.

TABLE 3

Element	Concentration wt%
O	42.24
Na	0.36
Mg	2.21
Al	11.26
Si	18.81
S	0.34
Cl	0.49
Ca	5.59
Ti	0.57
Mn	0.25
Fe	17.88
Total	100.00

Table 3. Results of the SEM-ED analysis. ZAF-correction has been applied.

Acknowledgements

We would like to thank Leila Krogh and her staff at the J.F. Willumsen Museum for giving permission to do the TL-analysis and for kindly providing help and assistance in the research of J.F. Willumsen's personal documents kept in the museum. We would also like to thank Benni A. Berg, the Department of Conservation, the National Museum of Denmark, for technical assistance in connection with the sampling procedure and Raymond Gwozdz for performing the neutron activation analysis. For reading the manuscript and offering helpful suggestions for improvements and changes we thank Bodil Bundgaard Rasmussen, the National Museum of Denmark and Leila Krogh, and for help with the translation Ashok Chaudhari. This work was supported by the Carlsberg Foundation.

NOTES

1. As for the exhibitions cf. the catalogues *Tanagra* 1996 and *Pegasus* 1997; For information on Willumsen, the 'Old Collection' and his antiquities see *Tanagra* 1996, 9-32, 89-160 and *Pegasus* 1997, 8-22. Concerning the fakes see *Tanagra* 1996, 40-44.

2. The inventory and other personal documents and letters of Willumsen referred to in this paper are kept in the J.F. Willumsen Museum Collection.

3. For Willumsen's purchases generally see *Tanagra* 1996, 33-39.

4. Examples of Attic black-figured phiala: Moore & Pease 1986, nos. 1427-1439, p. 56 note 1 for more references; Luschey 1939, 148, VII 1-8; Etruscan ones: Luschey 1939, 148, XIII.

5. Examples of the *omphalos* decorated with a Gorgoneion; Graef & Langlotz 1933, 112, no. 1249 (Attic black-figure); Sieveking & Hackl 1912, no. 995, Taf. 44 (Etruscan black-figure); Luschey 1939, 29.

6. Vian & Moore 1988, 253-254, 235 no. 298 (Apulian crater in Brussels).

7. Vian & Moore 1988, 259-260 (Poseidon with Nisyros), 255-256 (Athena spearing Encelados).

8. For Etruscan black-figure vases see Ginge 1987.

BIBLIOGRAPHY

Aitken, M.J. 1990
Science-based Dating in Archaeology. London.

Andrén, A. 1986
Deeds and misdeeds in classical art and antiquities (Studies in Mediterranean Archaeology 36) Jonsered.

Bothmer, D. von & J.V. Noble 1961
An Inquiry into the Forgery of the Etruscan Terracotta Warriors in the Metropolitan Museum of Art. (The Metropolitan Museum of Art Papers 11) New York.

Fake? 1990
Jones, M. (ed.), *Fake? The art of deception*. London.

Fakes 1973
Fakes and forgeries, The Minneapolis Institute of Arts. Exhibition catalogue (July 11 – September 29, 1973). Minneapolis.

Ginge, B. 1987
Ceramiche etrusche a figure nere (Materiali del Museo Archeologico Nazionale di Tarquinia XII) Roma.

Graef, B. & E. Langlotz 1933
Die Antiken Vasen von der Akropolis zu Athen II. Berlin.

Heydemann, H. 1876
Zeus im Gigantenkampf (Erstes Hallesches Winckelmannsprogramm). Halle.

Höckmann, U. 1991
Zeus besiegt Typhon, *AA*, Heft 1, 11-23.

Kolstrup, E. & V. Mejdahl 1986
Three frost wedge casts from Jutland (Denmark) and TL dating of their infill, *Boreas* 15, 311-321.

Luschey, H. 1939
Die Phiale. Bleicherode-am-Harz.

McKeever, S.W.S. 1985
Thermoluminescence of solids. Cambridge.

Moore, M.B. & M.Z. Pease 1986
Attic Black-Figured Pottery (The Athenian Agora XXIII) New Jersey.

Noble, J.V. 1965
The Techniques of Painted Attic Pottery. New York.

Pegasus 1997
Krogh, L. & P. G. Bilde (eds.), *Pegasus og Tanagra. Antikken i J.F. Willumsens kunst og samling*. Copenhagen.

Sieveking, J. & R. Hackl 1912
Die älteren nichtattischen. Vasen *Die königliche Vasensammlung zu München* I. München.

Tanagra 1996
Bilde, P.G. & L. Krogh (eds.), *Tanagra. J.F. Willumsen og hans antiksamling*. Aarhus.

Vian, F. 1952
La guerre des Géants. Le mythe avant l'époque hellénistique. Paris.

Vian, F. & M.B. Moore 1988
Gigantes, in: *LIMC* IV, 191-270.

Wiegand, Th. 1904
Die archaische Poros-Architektur der Akropolis zu Athen. Cassel & Leipzig.

Willumsen, J.F.W. 1947
Noter til "Gamle Samling". Copenhagen.

Kaare Lund Rasmussen
The National Museum of Denmark
Carbon-14 Dating Laboratory
Ny Vestergade 11
DK-1471 København K

Annette Gabrielsen Schmidt
Jægersborg Allé 239, 1. tv.
DK-2820 Gentofte

FORUM

THE ARCHAIC SETTLEMENT AT VROULIA ON RHODES AND IAN MORRIS

LONE WRIEDT SØRENSEN

Chapter 7 in Ian Morris' book, *Death-Ritual and Social Structure in Classical Antiquity* from 1992 is called "At the bottom of the graves: an example of analysis". It represents a re-evaluation of the archaeological finds from the Danish excavations of the Archaic necropolis and settlement at Vroulia on Rhodes (**Plate 15**) published by the excavator K.F. Kinch in 1914 and suggests a new interpretation of the site. Morris' declared intention (p. 179) is to find patterns. In the case of Vroulia he concludes that except for one group of graves, the adult cremation burials in the necropolis are structured according to descent and lineage, and that the child burials are located in between them without formal attachment to them. Furthermore, the settlement is interpreted as a rural habitation, and not a military garrison as suggested by Kinch. Morris' contribution is interesting and thought provoking, but a closer look at his analysis of the archaeological remains reveals that it is also problematic and calls for critical comments (c.f. Smith 1993), not least because the available data at times are treated casually. For the sake of clarity the present review of Morris' analysis follows his line of arguments through the material evidence from Vroulia.

Typology of the graves (**Fig. 1**)
Both Kinch and Morris divide the burials into seven groups, although they both doubt that groups I and II constitute separate units. Morris claims that he follows Kinch's group divisions, (p. 179), but he actually attributes some of the graves differently. In his Table 9 he adds grave 26 to group IV without explaining why, but seems to consider it part of group VI on p. 181. He is in doubt concerning grave 28, which is located north-east of grave 26, and grave 17 is attributed to group V. The in-

humation graves 18 and 30 are left out together with the cremated child in grave 29, because they deviate from the general pattern.

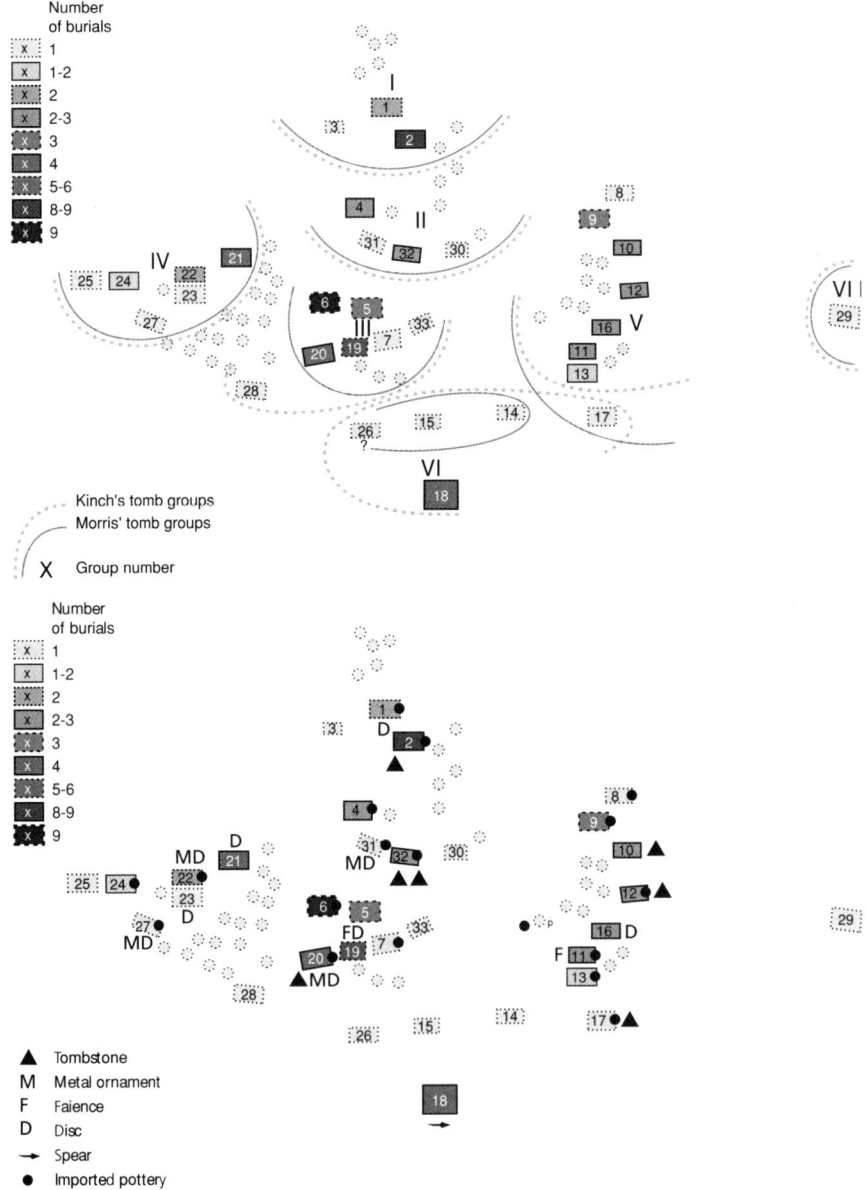

Fig. 1. Kinch's and Morris' grave groups and Kinch's estimation of the number of burials.

Kinch's grave groups (p. 35):

Grave group	Grave number
I	1, 2, 3
II	4, 30, 31, 32
III	5, 6, 7, 19, 20, 28, 33
IV	21, 22, 23, 24, 25, 27
V	8, 9, 10, 11, 12, 13, 16
VI	14, 15, 17, 18, 26
VII	29

Morris' grave groups (Table 9):

Grave group	Grave number
I	1, 2, 3
II	4, 31, 32
III	5, 6, 7, 19, 20, 33
IV	21, 22, 23, 24, 25, 26? 27
V	8, 9, 10, 11, 12, 13, 16, 17
VI	14, 15
VII	29
Uncertain	28

Demography

At first Morris also proposes an attribution of the infant inhumations to the various grave groups (Table 9). He assigns the infants buried between cremation group III and IV to group IV, although he expresses his doubts concerning which group they actually belong to (note 6). According to Kinch (p. 37) five main types of pots were used as urns. Morris (p. 181) mentions four types but lists five types. However, he finds that the main distinction is between the amphorae and the others "which are all ordinary household vessels". Furthermore, he reclassifies some of the pots. Like Kinch, Morris divides Kinch's group I, *cythrai*, into two groups; pots with one handle, i.e. jugs, and pots with two handles, i.e. cooking pots. However Morris classifies pot *v* and *kk* as cooking pots, although they are described in the publication as having one handle. Morris also classifies *k* from Kinch's group II as a jug, and *jj* and *pp* also from Kinch's group II are added to the cooking pots

245

together with some of the pieces from Kinch's group III. Pot *f* is listed together with the amphorae and the Cypriot-looking amphora *gg* is left out.

Kinch considered five of the amphorae from his group IV (*e, j, r, z, qq*) close parallels, but only four of them are listed as SOS type amphorae by Morris. The two lists of hydriae correspond, except for *ii*, which is missing from Morris' list. According to Kinch his group I together with *gg* was used for burials of the smallest infants, because the bodies were placed through the opening of the vessels. In fig. 42 Morris lists the shapes according to size, but it should be noticed that Kinch only gives exact measurements for jug *ff* (29.5 cm), pots *jj* (36 cm), *cc* (39 cm) and *aa* (42 cm), which belong to Kinch's group II and III and hydria *x* (42.5 cm). The remaining hydriae measure about 46 cm, and the amphorae including *f*, range from 51 to 99.5 cm. Amphora *gg* is left out. Morris' rearrangement of the amphorae makes sense if *gg* is omitted, and according to the information available they represent the tallest shapes. Although hydriae are best represented in the graves located between burial group III and IV, the vessel types show a comparatively even distribution. However, Morris' exercise (p. 182) serves to document a clear correlation between the pot type and size and to illustrate that the use of grave goods and types of burial vessels seem to underline age groups rather than dividing children on other principles within the age groups.

Kinch's classification:

I: cythra à une anse	c, g, h, i, v, ff, kk
cythra à deux anses	a, b, l, q, hh, ll, beta
II: jarre de forme plus élancé	k, jj, pp
III: jarre de sorte d'amphore, à pense arrondie en bas	t, cc
jarre de sorte d'amphore à pied annulaire	f, aa, gg
IV: amphores	d, e, j, m, p, r, s, z, bb, qq
V: hydries	o, u, x, y, dd, ee, ii, mm

Morris' classification:

1: jug	c, g, h, i, k, ff
2: cooking pot	a, b, l, q, t, v, aa, cc, hh, jj, kk, ll, pp
3: amphora	d, f, m, p, s, bb, qq
4: SOS type amphora	e, j, r, z
5: hydria	o, u, x, y, dd, ee, mm
6: ?	n, gg, nn, oo

Morris does not state the exact number of burials he uses for his calculations concerning the proportion of children and adults in each burial group. Here he seems to follow Kinch's estimation (p. 89) of about 125 burials, 43 of which represent small children, including grave 29. However, it is more difficult to assess the number and age of the cremated individuals. Apparently it was sometimes difficult for Kinch to estimate the exact number of burials deposited in the same grave, and they comprise individuals down to the age of six. As it appears from the list below, the total number of cremation burials ranges according to Kinch from a minimum of 67 to a maximum of 78. To these the inhumations in graves 18 and 30, counting five male adults, should be added.

List of graves according to possible number of burials:

1 burial	Graves 3 (child), 7, 14, 15? 17? 23, 25, 26, 27, 28, 31? 33 (14-16 years old)
1-2 burials	Graves 8, 13, 24
2 burials	Graves 1, 22
2-3 burials	Graves 4, 10, 11, 12, 16, 32
3 burials	Graves 5? 9?
4 burials	Graves 20? 21?
5-6 burials	Grave 19
8-9 burials	Grave 2
9 burials	Grave 6?

The grave goods (Fig. 2)

Seven of the pot burials contained gifts (*f, m, p, s, bb, ff, qq*), and as observed by Morris they do not form any particular pattern according to their location in the necropolis. Six of them contained from one to six objects, but grave *s* held three small deposits of altogether 17 objects.

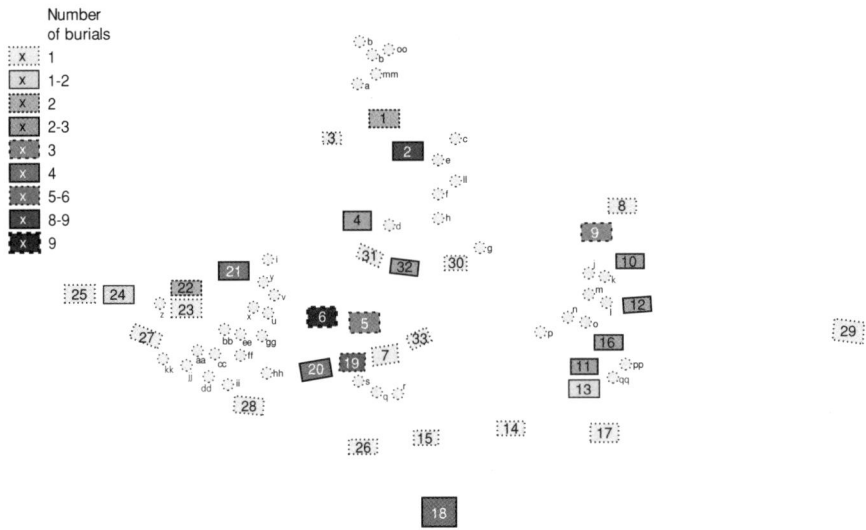

Fig. 2. Graves with "exotic" finds.

According to Morris (p. 182 quoted above) the use of grave goods also underlines age boundaries. It should be noticed though, that the burial vessel in grave *ff*, for instance, is a jug, presumably used for the smallest children and that gifts were not found in association with the larger burial vessels such as hydriae and several of the amphorae.

In his Table 10, listing grave goods associated with primary and secondary adult cremations, Morris includes grave 3, which according to Kinch contained a child about six years old, but leaves out graves 1, 10, 12, 13, 15, 16, 20 and 32, which also contained gifts. It is furthermore difficult to understand why Morris assigns 32 objects to the primary deposit and 18 to the secondary deposits in grave 2. Here only the first eight items are listed with indications of depth below surface. Furthermore, Kinch (p. 59) states that it was not possible to distinguish between the different layers. Concerning grave 11 Morris assigns four pots to the secondary deposits, while Kinch (p. 70), having expressed doubts that they actually belong to the grave, states that he does not know which level they come from.

Morris (p. 185) mentions that six of the so-called discs were associated with primary burials and 12 with secondary burials. In his Table 10, seven are listed under primary burials, and to these should perhaps be added one of the five, not four discs, found in grave 22. Morris omits four other discs found in graves 1, 16 and 20, three of which seem to belong to primary deposits. The deposition of discs in graves, which is also known from other places in Rhodes (Friis Johansen 1957, 187) is connected only with the cremation tombs. It is interesting to note that at Vroulia nine of the altogether 24 discs were found in graves of group IV, which Morris is keen on interpreting as a special burial unit. He is right that some of these tombs are just as rich as tombs belonging to the other groups, but it is perhaps even more interesting that the same types of objects were deposited as grave gifts here as in the other graves including the discs. It should be mentioned that Morris' Table 12, listing imported finds, also contains incorrect information. For instance, grave 1 contained three and not two Corinthian pots, grave 2 contained 21 imported pots, grave 20 contained a Proto Corinthian lekythos not listed, and grave 26 contained five, not three fibulae.

Grave size
Morris pays great attention to the depth of the cremation graves, and agrees with Kinch that "it was determined by the number of burials intended to be deposited in them, *before* the first cremation took place" (p. 190). The terrain slopes towards the north east, and the depth of each grave varies according to Kinch (p. 53) almost everywhere, being 0.20 to 0.35 m deeper towards the west. In grave 8 and grave 10 the difference is even 0.40 m. It would have been fair of Morris to inform his readers that in cases where a tomb is provided with two depth measurements, he has calculated an average depth. Kinch lists two measurements in connection with graves 8, 9, 10, 16-25, 27, 31-32. Graves 12, 26, 28, 33 were measured in the centre, and the depth of one of the corners is indicated for graves 1, 13, 18, 55. For the remaining seven graves Kinch only provides the measurement, not where it was taken (p. 55 f). However, four of the graves of group IV with or without grave 26, are indeed shallower than the others, but not graves 21 and 23, or 26 for that matter. According to Morris (p. 192) those who buried their dead in group IV, which is now described as "being separated from the main cemetery by a band of child graves, did so with a very different attitude towards the symbolic expression of descent". Some of the other graves, such as graves 7, 14, 15 and 17, are not much deeper, and it is highly speculative to describe group IV as being separated by the child tombs

located between group III and IV. The space between group III and IV is actually not greater than that between some of the other groups. The space between group I to III and V may even have been more conspicuous, if a road passed between them as suggested by Kinch (p. 36). Even if we accept that those buried in tombs of group IV were somehow outsiders, four of them were buried in grave 21 and perhaps two in graves 22 and 24. If descent or lineage procedure was not followed here, we are dealing with what might be termed collective descentless graves, to follow Morris' terminology.

Morris (p. 193) emphasises the significance of the tombstones, which according to him were found over some of the deepest graves destined to hold several individuals. Kinch mentions (p. 55) that tombstones were found over graves 2, 10, 11, 12, 17 and 20. However, according to his description of the graves, no tombstone was associated with grave 11, while grave 32 was provided with not one, but two tombstones, as also indicated on Morris' fig. 46. Here it should be noticed that grave 17 with a depth varying from 0.35 to 0.50 m contained only one burial, and no markers were associated with deep tombs like graves 6 and 19, both containing several burials. Based upon this and the small number of extant tombstones, the significance of their absence from burials of group IV appears to be less noteworthy than indicated by Morris. However, Morris concludes "that group IV deviates from the descent dogma of the Vroulian ritual", and he proposes that "the unstructured and descentless graves in this group could be mortalities from visiting ships". According to him this seems to be supported by the fact that group IV only contained 10 cremations. Compared with the number of cremations in the other groups this is not necessarily significant and depends upon how one counts.

Number of cremations in tomb groups:

group I	Minimum 10	maximum 11
group II	Minimum 6	maximum 8
group III	Minimum 22	maximum 24
group IV	Minimum 10	maximum 11
group V	Minimum 13	maximum 19

Based upon the above, neither the location of the graves, their depth and reuse, the lack of tombstones nor the distribution and types of grave goods, and in particular the discs speak in favour of Morris' interpretation. One would furthermore expect that in those days people who died at sea were disposed of *en route* in order to avoid freighting rotting bodies. But, if the group IV graves were indeed used for dead foreigners, it is interesting to observe that they were treated very much like the local deceased.

The settlement
The settlement consisted of at least two rows of houses inside a wall. Kinch (p. 123) distinguished three types of rooms and interpreted 13 rooms without doors as store rooms. Kinch (p. 116) also suggested that several rooms formed separate units and mentioned in particular I 6-7 and I 25-26. Morris (p. 193f) follows the suggestion and proposes altogether 15 to 20 units, a number which fits nicely with the approximately 80 cremations, which according to him represent about 40 adults burying two generations of their dead. However, Morris only ventures to specify some of these house units, i.e. I 18-22, I 23-26, I 27-29 and I 29-32, probably because he also finds it difficult to separate one house from the other (cf. Melander 1988, 85). It should be mentioned that incorrect information is also provided in Morris' survey of the settlement. Morris claims that room I 22 is not provided with a door, which is contradicted by Kinch's statement and the published drawing of this particular room (Kinch p. 114, fig. 35). The room has an *ante*room with doors, and it is difficult to see Morris' stub of wall, which he suggests formed part of a courtyard. Furthermore, room I 32 and not room I 18 is the second richest source of pottery in the settlement, room II 4 with a slab floor is interpreted as a store room, and to call room I 32 "a much grander building" is an exaggeration.

Concerning the finds in the rooms Kinch suggested that room I 2 and I 32, which contained most pottery finds, may have belonged to pottery dealers. Morris rightly states that people would have taken their valuables with them and that conclusions based upon the remains are questionable. It should be noticed though, that various types of decorated cups clearly predominate among the pottery, followed by oinochoae. This shows that the inhabitants were able to acquire fine drinking vessels, or that decorated pottery was not very valuable.

The interpretation of Vroulia
Morris rightly states that the settlement plan is only relevant to the period immediately before Vroulia was left, but there is no way of telling if and how much change took place during the period of occupation. Based upon the regularity of the site and what he considered the defensive wall and a tower, Kinch (1914, 5) interpreted Vroulia as a military outpost on the southern tip of Rhodes *en route* for ships heading across the Aegean. A defensive character of the wall is furthermore supported, if Melander's suggestion (1988, 85) concerning the gate is correct. Morris, on the other hand (p. 187), referring to the slight structure of the wall and the age structure of the cemetery, finds that "it would fit a "normal" agricultural population far better than a putative garrison". Whether the intention of the wall was defensive or not, it may at least have served to keep out – not Morris' wild animals, but flocks of sheep and goats grazing in the rather barren landscape surrounding Vroulia. Morris does not take into account that his redefinition of the wall provokes a redefinition of Kinch's tower at the gate. Following Morris' interpretation of Vroulia as an agricultural settlement, a function as a communal storage facility might inter alia be suggested.

However, "as the terrain governed the line of the wall, and the rows of excavated houses in their turn were governed by the wall" (Melander 1988, 83), the location of Vroulia and the surrounding landscape are important parameters for interpreting the settlement. It is isolated, and access from the sea ranges from difficult to impossible. More importantly, the surrounding landscape is definitely not a farmer's dreamland. Unless drastic changes have taken place since antiquity which is of course possible – nothing much could have been grown in the area described by Morris (p. 198) as an ideal peasant world, preserving the proper stable relations between men in spite of the disruptive forces of birth, marriage and death.

Vroulia is indeed a fascinating place, but "although a unique site – like a unique grave – is open to any number of direct interpretation" (Morris p. 193), his interpretation of Vroulia is based on a questionable treatment of the archaeological evidence from the site.

BIBLIOGRAPHY

Friis Johansen, K. 1957
Exochi, ein frührhodisches Gräberfeld, *Acta Archaologica 28*, 1-192.

Kinch, K.F. 1914
Fouilles de Vroulia (Rhodes). Berlin.

Melander, T. 1988
Vroulia: Town Plan and Gate in S. Dietz and I. Papachristodoulou, *Archaeology in the Dodecanese*, The National Museum of Denmark, Copenhagen, 83-87.

Smith, Chr. J. 1993
Review of Ian Morris, Death-ritual and social stucture in Classical antiquity, 1992, *Antiquity* 67, no 255, 464-65.

Lone Wriedt Sørensen
Department of Archaeology and Ethnology
University of Copenhagen
Vandkunsten 5
DK-1467 Copenhagen K

lws@hum.ku.dk

CURRENT DANISH CLASSICAL ARCHAECOLOGICAL FIELDWORK

ANNETTE RATHJE (ed)

With contributions by Søren Dietz & Sanne Houby-Nielsen, Klavs Randsborg, Lone Wriedt Sørensen, Maria Berg Briese, Anne Marie Carstens & Poul Pedersen, Pia Guldager Bilde, Christina Trier

GREECE
Greek-Danish excavations in Aetolian Chalkis (1995-2000)
(Søren Dietz & Sanne Houby-Nielsen) *Plates 16-20*
Directors: Søren Dietz (Ny Carlsberg Glyptotek), Copenhagen and Lazaros Kolonas (6th Ephoria of Prehistoric and Classical Archaeology, Patras). *Field directors*: Sanne Houby Nielsen (University of Lund, Sweden) and Ioannis Moschos (6th Ephoria, Patras).
Financial support: Consul General Gösta Enbom's Foundation
Goal: The aim of the project was to identify the site of ancient Chalkis in Aitolia, mentioned by Thucydides, and to study the topography and architectural features of a Corinthian stronghold along the coast of the Gulf of Corinth. Furthermore, the goal was to study the economic, social and religious characteristics from a historical perspective and, in doing so, to estimate the chronological development of the stronghold from the Bronze Age – the site was mentioned in the Catalogue of Ships – to the habitation in Hellenistic times.
Result: Since 1995 a Greek-Danish archaeological project has been under way in the area centred upon the fishing village of Kato Vassiliki. This village is built around a small natural harbour on the Aetolian coast, directly opposite Patras. The project is being carried out under the auspices of the 6th Ephoria of Prehistoric and Classical Archaeology in Patras and the Danish Institute at Athens.

The harbour of Kato Vassiliki is framed by two large mountains that in Antiquity were called Chalkis (to the west) and Taphiassos (the one to the east): today they are known as Mount Varassova (914 metres)

and Mount Klokova (1037 metres) respectively. Two sites within this area constitute the focus of the project: the extensive fortified site on the eastern slope of Mount Varassova that is called Pangali, and the low hill at the eastern end of the bay, known as Haghia Triadha because of its huge Early Byzantine basilica. This place has a long tradition of wealth in ancient pottery.

An intensive survey carried out in 1995 proved that the site of Pangali had served merely as a kind of military refuge or stronghold and that it had not been used for long-term settlement. In sharp contrast to this, the hill of Haghia Triadha could, on account of the richness of the material discovered, safely be identified as Aetolian Chalkis, as mentioned by such writers as Homer, Thucydides, Polybius, and Strabo. The excavations since 1995 have, accordingly, mainly been concentrated on the hill of Haghia Triadha, with the exception of a trial excavation beneath a cave-like structure at Pangali, which led to the discovery of an enormous concentration of Final Neolithic pottery and the remains of a fireplace.

Seven trial trenches (ranging in length from 2-80 metres) and 8 large excavation units (generally 5 x 5 metres) were dug in the summer of 1996-1998, in most places down to bedrock (**Plate 16**). In addition, a considerable amount of work has been put into the cleaning, excavation and measuring of the Byzantine wall which encircles the hill and which has now been more accurately dated to the Early Byzantine period, contemporary with the basilica. It should also be mentioned that the geologist Kaj Strand Pedersen has, among other things, demonstrated the existence of marine layers immediately to the west of the hill, with lower strata dating back to the Early and Middle Helladic period.

At the present stage of the excavation, it is certain that a large-scale settlement began on the hill of Haghia Triadha in the Late Neolithic/Early Helladic period, extending down the western slope in particular towards the area of marine layers. This settlement should probably be regarded as a continuation of the Final Neolithic site on Pangali. It is therefore likely to provide important new information on settlement patterns and movements in this period. Judging by the many finds from the Late Bronze Age, Geometric(?) and Archaic periods, discovered in floating layers in most of the trenches, the hill was inhabited throughout these periods as well.

Particularly noteworthy, is the rich 7th century material found in situ in trench 4 on the western slope. This indicates that the area occupied by Chalkis in this crucial period was quite extensive. During the summer 2000, excavations continued in the area around trench 4 in the lowlands west of Aghia Triada. Two well-defined horizons with foundations for Archaic buildings were found, dating back to the 7th and to the 6th century BC respectively (**Plate 17**). Approximately 30 cm below the lower Archaic levels, the foundations for a large Mycenaean (LHI-IIC) *megaron* were uncovered.

The so-called "Acropolis Wall", a huge wall built from large blocks of sandstone, is presumed to have been erected in the 5th century. It encircles the upper plateau of the hill, and remains of Classical houses – one of which possessed a yard with storage facilities – have been found both inside and outside this wall. The house remains in trenches 1-2 in **Fig. 1** point to an orthogonal city plan. Moreover, judging from the fragments of fine terracotta roof-decoration stemming from the upper plateau, one or more Classical temples once crowned the hill, presumably on the site of the present basilica.

In the Hellenistic period, a series of houses were built immediately outside the "Acropolis Wall". Their walls were simple, dry-stone walls, but carried a fine terracotta roof in the Laconian manner (**Plate 18**). A wealth of bronze coins, primarily Aetolian League coins, as well as table- and household ware were discovered in the houses. In the plain to the west of the hill, a large Hellenistic Cist grave was excavated during the 1999 campaign (**Plate 19**). A skeleton was placed in the middle of the grave on its back, while the bones of a further three people were piled in the corner. More than sixty vases, a bronze mirror, and various terracottas were included in the finds that were made in the grave (**Plate 20**).

Bibliography: ArchRep 1996-1997, 43; *ArchRep* 1997-1998, 43-44; Søren Dietz, Lazaros Kolonas, Ioannis Moschos and Sanne Houby-Nielsen (eds.), Surveys and Excavations in Chalkis, Aetolias 1995-1996. First preliminary report, *Proceedings of the Danish Institute at Athens II*, 1998, 233-315; eadem (eds.), The Greek-Danish Excavations in Aetolian Chalkis, 1997-1998. Second preliminary report, *Proceedings of the Danish Institute at Athens III*, 2000, 215-301.

40. Kephallénia – The four Poleis
(Klavs Randsborg)
Cf. *ActaHyp* 5, 1993, 394-99

Ancient Kephallénia, the largest of the Ionian Islands (nearly 800 km²), held four ancient Greek poleis, Pale (near Lixouri), Krane (at Argostoli, the modern capital), Same (today's Sami), and Pronnoi (with an inland city centre, later supported by coastal Poros), each occupying a distinct part of the island (**Fig. 1**). Nearby Ithaka (100 km²) made up the fifth polis of the area. The landscapes of Kephallénia (and Ithaki) are dramatic and very beautiful with fine conditions for husbandry and agriculture (precipitation is high) in several places, including the lowlands, and excellent natural harbours. The high mountains (1630 m) of eastern Kephallénia still carry coniferous forests. In spite of the naval rôle and important strategic position of Kephallénia at the mouth of the Bays of Patras and Corinth, only little is known about its ancient history, although the presence of Athens and of the Macedonian powers was certainly felt; Rome conquered the islands in 189/188 BC.

The Project. Under the auspices of the 6th Ephorate for Antiquities, Patras, an archaeological investigation of the landscapes and sites of northern, central, eastern, and south-eastern Kephallénia, was carried out during the early to mid-1990s in the main by Institute of Archaeology and Ethnology at the University of Copenhagen. Common methodology stresses highly detailed cross-country surveys in limited areas. However, the huge Kephallénian operation has been a conscious mixture of intensive and extensive work. Aided by detailed artefact studies, this has provided an overall picture of the archaeology of all periods across indeed very large areas, and thus enabled a cultural, geographical and historical approach to the development of settlement and monuments. The pioneer mapping and detailed typological-chronological study of the ancient walls, almost stone by stone, has been very detailed due to the unusually complex nature of the data, including many unfinished and highly ruined walls (of cities and structures alike) in 27 different types of masonry. The archaeological surveys have revealed a total of 457 sites representing each main chronological phase from the Palaeolithic to c1500 AD (the city centres comprise several settlement units each) (**Fig. 2**).

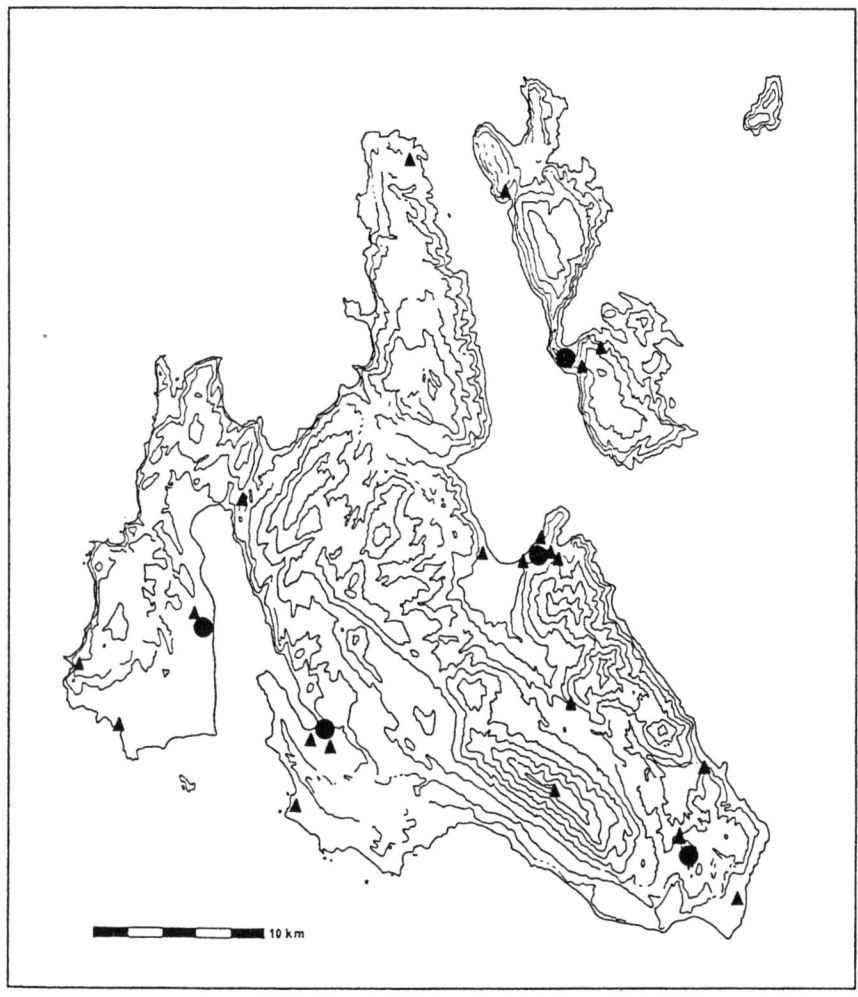

Fig. 1. Kephallénia-Ithaka. Ancient poleis centres (large dot) and ancient temples and shrines (triangle; Archaic period onwards).

Prehistory and Early History. The Stone Age settlement of Kephallénia is extensive, including major Middle Palaeolithic coastal sites, a large quarrying site, and many Neolithic and Early Bronze Age settlements. From the Middle and Late Bronze Age come numerous both settlement sites and graves, including fine tholoi and several fortified sites. The Iron Age, including the Geometric and Archaic periods, sees the establishment of several sanctuaries both at the city centres, at the borders between poleis, and at prominent sites in the landscape, as well as

	PRONNOI TERRITORY			SAME TERRITORY		
	City	Chora	*Sum*	City	Chora	*Sum*
(Middle) Palaeolithic	0	8	*8*	0	2	*2*
Neolithic	1	6	*7*	0	3	*3*
(Early) Bronze Age	1	2	*3*	0	0	*0*
Prehistoric wares	8=28%	21	*29*	6=30%	14	*20*
"Mycenean Red" wares	6=24%	19	*25*	1=10%	9[1]	*10*
(Sub-Mycenean/) Geometric	4	1	*5*	1	0	*1*
Archaic	2	3	*5*	11=79%	3[1]	*14*
Late Archaic/ Early Classical	4	4	*8*	8=73%	3[2]	*11*
Classical/Late Classical	6=24%	19	*25*	21=81%	5[1]	*26*
Early Hellenistic/ Hellenistic	10=53%	9	*19*	28=67%	14[2]	*42*
Late Hellenistic	0	6	*6*	21=68%	10	*31*
Imperial Roman	6=19%	25	*31*	15=36%	27[1]	*42*
Late Antiquity	1	5	*6*	6=35%	11	*17*
(High) Middle Ages	16=41%	23	*39*	11=48%	12	*23*

Fig. 2. The number of surveyed archaeological site units with lithics (top), sherds etc. (middle), or sherds/roof tiles (bottom) indicative of the named narrow phases at, respectively, the city and the chora of the (later/former) poleis of Pronnoi and Same, eastern Kephallénia. The percentages, indicative of the relative weight of the settlement on the city, should be read with a standard deviation of at least five to ten points due to the small numbers of sites. (The Mycenean fortress/frontier sanctuary of Digaleto is added to Same, the number of extra site units is given in brackets but not included in the percentages.)

a series of small round or square border fortresses. The poleis centres themselves were seemingly not fortified until the (later) fifth century BC. Incidentally, Corinth does not seem to have played a strong role in Kephallénia (contrary to Ithaka), although it may have provided the stimulus for several sixth century BC temples.

Classical Antiquity. Apart from the city centre of Krane and a very few remains at Pale, etc., ancient walls only seem to have been preserved in the eastern parts of Kephallénia. Planned walled cities, in various states of completion, have been found at Same, Pronnoi, Poros, Krane, plus (possibly) Aëtos (Ithaka). Smaller architectural sites, fortresses, towers, temples, etc., are also known. While the Kephallénian poleis have a deep archaeological history of their own, some even going back to the Neolithic, the enceintes and elegant planned city- and townscapes are seemingly the result of "foreign" intervention. Athens was the major external power on Kephallénia during the mid- to late fifth, and, especially, the early fourth centuries BC (the "Messenian" fortification at Krane of the late fifth century BC; the elegant planned city and impressive parts of the enceinte of Same; various large fortresses, etc., all from c375 BC). Around 300 BC the major external powers were the Macedonian kingdoms. Likely, a Macedonian power was responsible for a fine tholos temple (**Fig. 3**), most of the enceinte of Same, which is in Eastern Greek masonry, and, in particular, for the impressive bastioned (but unfinished) enceinte near Krane with its large Athenian style dipylon gate, behind which is a truly huge, but never built, planned metropolis, known only from the first grids of its intended streets (**Fig. 4**). "New Krane" was probably connected with the western aspirations of Demetrios "Poliorketes", which may also explain why the metropolis was never finished (290/289 BC). The Aitolian league was felt around 200 BC, and Rome, of course, very much thereafter. The archaeological surveys have revealed a particularly rich settlement in the (later) Classical and Early Hellenistic periods (before 200/150 BC). A change in the general pattern, from numerous and generally higher lying sites in the "Greek" to fewer and lower lying ones in the "Roman" period – including several fine coastal villas of the imperial centuries – is also noted. Furthermore, for the "Greek" period, Same displays relatively little settlement outside its fine city, while, by contrast, the territory of neighbouring Pronnoi sees many rural settlements and only a small city (although with the harbour town of Poros). Seemingly, no common settlement is situated higher than the shrine on the acropolis of the polis centre.

Middle-ages to recent times. Late Antiquity, with an important Christian basilica from around 500 AD at coastal Panormos/later Phiscardo

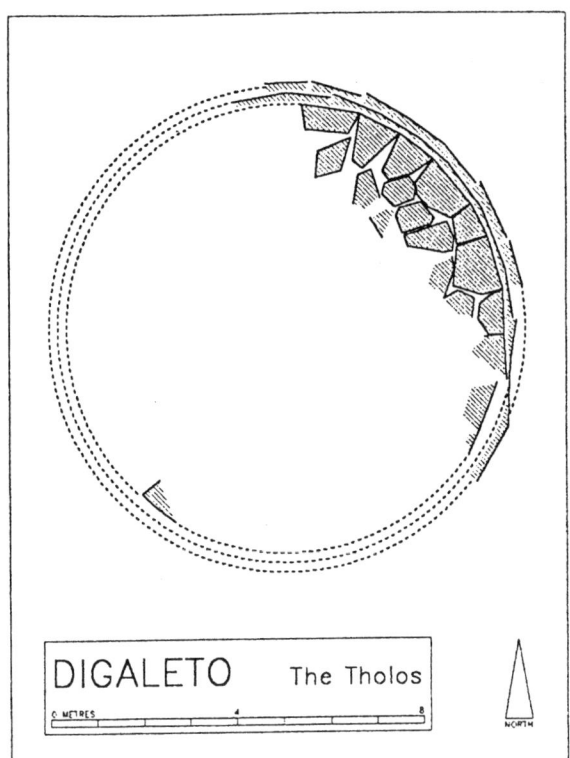

Fig. 3. Digaleto (to the South of Sami, between the poleis cities of ancient Same and Pronnoi, eastern Kephallénia). Tholos temple from c. 300 BC (incidentally set in an ancient shrine, in turn in a Bronze Age fortress; cf. Fig. 2).

(to the south of Nicopolis, at present-day Preveza) marks a new growth in settlement. Byzantine Kephallénia is relatively poorly known, but Kephallénia did give name to a military district (*thema*). In the Norman/Italian phase (c.1100/1200 to 1500), the acropoleis of the ancient cities were re-fortified to serve as strongholds (and estate centres) and several monasteries built. Further building, including fortresses, manors and fine churches, took place in the Venetian period (till c.1800 AD). The British (till 1864) in particular developed the towns of western Kephallénia, and built roads and bridges. The population of Kephallénia peaked in the early twentieth century AD, while in 1953 a major earthquake destroyed almost all of Kephallénia. With communication already then being more important than optimal location vis-à-vis the agricultural and highland resources, the new and less populous settlement moved down to the main roads. This left a Greece before modernity in the form of dozens and dozens of silent ghost-villages, an ethnological present (with informants still alive) turning slowly into archaeological data.

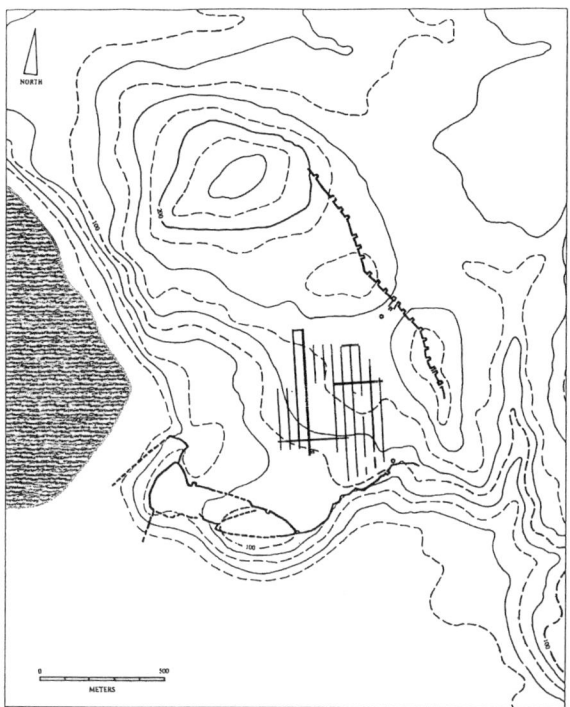

Fig. 4. The ancient City of Krane (to the East of Argostoli, western Kephallénia), at one of the best natural harbours in the Mediterranean, with the planned extension of "New Krane" to the East and North (of c. 300/290-289? BC).

Conclusions. Kephallénia (with Ithaki) holds a rather anonymous position in the written history of Greece and was always marginal to the Mediterranean development. The archaeology of Kephallénia, however, has proved much richer than anticipated from earlier work on the island; most of the information acquired by the present project is hitherto unknown. Indeed, Kephallénia is a fine example of the interplay betweeen the "little history" of local entities, materially defined and bound together in a regional patchwork, and the major forces, setting the larger stage, as well as being culturally dominant. Local society may seem a victim to such forces, but the latter never succeeded for very long. This is one of the lessons of traditional Mediterranean civilization. The contrast with the North-European homeland of the visiting archaeologists on Kephallénia is striking – steady growth millennia by millennia.

Bibliography: K. Randsborg (ed.), *Kephallénia. Archaeology & History. The Ancient Greek Cities.* Acta Archaeologica Supplementa IV (double vol.) = Acta Archaeologica Vol. 73, 2002 (forthcoming).

CYPRUS
9. Excavations at Panayia Ematousa, Aradippou 1995-1999
(Lone Wriedt Sørensen)
Cf. *ActaHyp* 4, 1992, 380-381; 5, 1993, 404; 6, 1995, 306-309.
Direction: Lone Wriedt Sørensen, Copenhagen.
Financial Support: University of Copenhagen; the Danish Research Councils; the municipality of Aradippou, Cyprus.

Five more excavation campaigns, each lasting four weeks, were carried out at the rural habitation in the vicinity of the small church of Panayia Ematousa near the village of Aradippou in Cyprus. So far, a settlement convering an area of little less than 1000 square meters has been exposed, but the habitation clearly extends beyond the excavated area (**Plate 21**).

It has been underlined that several phases of occupation are represented on the site. Apart from a few Bronze Age sherds, the earliest substantial amount of pottery, which dates to the Archaic to early Classical period, was found scattered in the excavation with concentrations appearing in pits dug into the subsurface. The material found in the pits, including stone objects and fragments of figurines, is consistent and seems to have been discarded simultaneously. Moreover, during the later excavation sessions architectural remains belonging to this period were revealed. They consist of a series of rooms situated in the north-western and southern part of the excavated area. The building technique of the walls and their orientation differs slightly from the walls belonging to the later phases, but in some cases they were reused in the construction of later walls.

According to the results obtained so far, the main occupation of the habitation dates to the late Hellenistic to early Roman period. During this phase, the ground plan is characterised by long walls running approximately north – south and seems to follow a general lay out. Most of the rooms are fairly small. Some of them contain benches, others cists or basins constructed of large stone slabs. The material found in the rooms indicates that they were used for agriculture production and storage. Several stone built "cellars" were probably also used for storage. Despite the effort to define certain limits to the settlement, no streets were found, and as only some of the rooms are provided with doorways it is difficult to subdivide the architecture into well-defined units. However, a series of rooms seem to connect with a courtyard in the centre of the excavation, and in the eastern part of the excavation removal of the baulks also revealed that a series of smaller rooms interconnected by doors lead to a larger room to the south, which measures approximately 5 x 7 m.

In the centre of the excavation remains of substantial walls, which belong to either the early phase or an intermediate phase were also re-used during this period. The area south of these walls, which had been filled up with large ashlar blocks and slabs, was tentatively interpreted as an open area. The large stone slabs lying in the eastern part were removed, and a type of feature, not documented before, was exposed. It consists of a groove cut into the bedrock. It runs roughly north – south and is little more than 3 m. long and 0.10 m. wide. At the bottom are 5 circular depressions with equal space in between them giving an initial impression is that a fence was once secured here screening off part of the area before it was filled up.

Material dating from the third to the fourth century AD is scarce suggesting that the settlement may have been abandoned for a period of time. During the late Roman period the site was reoccupied, but as the architectural remains are lying just below the surface soil the evidence has largely been ploughed away, and only a few walls and floor slabs remain in situ. However, the finds indicate that the main period of this late phase of habitation dates from the sixth to the seventh century.

The excavation has also revealed that the formation of the bedrock is highly irregular. Part of the hill, where the settlement is located, is actually formed by accumulation of human building activities and thus represents what we may call a pseudo tell.

Following the policy of the Department of Antiquities in Cyprus, the finds and results of the investigations so far obtained are now being prepared for publication. However, as the settlement covers a larger area and raises many questions it deserves further investigation. Evidence of habitation during the periods in question is rare and confined to the city centres. The settlement at Panayia Ematousa represents so far the only example of a rural habitation in Cyprus, and the focus of current field projects in Cyprus on identifying similar types of sites reflects their importance.

Bibliography: L. Wriedt Sørensen in: D. Christou, Chronique des fouilles et découvertes à Chypre en 1993, *BCH* 118, 1994, 679; 1994,

BCH 119, 1995, 824; 1995, *BCH* 120, 1996, 1070; 1996, *BCH* 121, 1997, 905; L. Wriedt Sørensen with contribution by J. Lund, Preliminary Report of the Danish Archaeological Excavation to Panayia Ematousa, Aradippou 1993 and 1994, *RDAC* 1996, 135-157; L. Wriedt Sørensen, Aradippou Panayia Ematousa in: E. Herscher, Archaeology in Cyprus, *AJA* 102, 1998, 334-5; L. Wriedt Sørensen with contributions by A. Destrooper-Georgiades, R. Frederiksen, K. Winther Jacobsen and J. Lund, Third Preliminary Report of The Danish Archaeological Excavations at Panayia Ematousa, Aradippou, Cyprus, *PDIA* II, 1998, 319-381.

TURKEY
37. Survey and excavations in Halikarnassos, Turkey (1995-2001)
(Maria Berg Briese, Anne Marie Carstens & Poul Pedersen)
Cf. *ActaHyp* 4, 1992, 378-379; 5, 1993, 399-404; 6, 1995, 309.

Direction: P. Pedersen in collaboration with the Directorate of Bodrum Museum. A. M. Carstens acted as field director of the citywall excavations and M. B. Briese is in charge of the registration of finds and the pottery studies. S. Isager is epigraphist of the project and B.Berg directs the conservation and restoration works. B. Poulsen is in charge of the "Late Antique *Domus*" project and C. Briese was in charge of the rescue excavation of rock-cut tombs north of the Myndos Gate. In 2000 and 2001 a team directed by K. Jeppesen studied the pottery finds from the Maussolleion excavations of 1966-77.
Financial support: University of Southern Denmark, Ericsson Türkiye and Türkcell, The Danish Research Council for the Humanities, The Novo Nordisk Foundation, The N.M. Knudsen Foundation, The Svend G. Fiedler Foundation, The Carlsberg Foundation (for the Maussolleion Project).
Goals: The numerous different works carried out in relation to the Halikarnassos Project are based on two main aims:
A. The discovery and study of monuments that can illuminate the role of Halikarnassos in the "Ionic Renaissance" of the fourth and early third century BC.
B. The discovery and study of remains of all ancient periods that can illuminate the cultural, commercial and art-historical place and identity of Halikarnassos in the Ancient world.

In the period 1995-2001 these aims have been pursued by: 1. Architectural studies, – in particular the fourth century BC fortifications of Halikarnassos; 2.Studies related to the Salmakis remains and its important inscription; 3. Registration of finds; 4. A Ph.D. project on the Tombs of the Halikarnassos peninsula; and 5. A Ph.D. project on the pottery of Halikarnassos.

1. Architectural studies
The City Wall Project
In 1998 the Turkish authorities decided to initiate a large-scale restoration project of the city wall of Halikarnassos, sponsored by the Turkish companies Ericsson Türkiye and Türkcell. The Danish Halikarnassos Project was invited to carry out architectural and archaeological studies in relation to the restoration project also sponsored by the mentioned Turkish companies. These studies were carried out as surveys and excavations in three campaigns of 1998, 1999 and 2000. The main results were: The discovery at several places of the fortification ditch mentioned in ancient written sources, the discovery and re-discovery of three gates, the location of stairs leading to the wall-walk, the discovery of new towers in addition to those already known, the observation of several cases of "indented trace" and the identification of specific technical and constructional features, by which the Halikarnassos fortifications can be related to other important fortifications of the time of Maussollos and Alexander. The survey led to the conclusion that ancient Halikarnassos had at least four strong fortresses integrated in the city wall circuit.

The documentation of ancient building stones reused in the Turkish houses of modern Bodrum gave new evidence concerning the late-classical Türkkuyusu temple and monumental public buildings of early Hellenistic date, – possibly those mentioned in the well-known building inscriptions of Halikarnassos.

2. The Salmakis Fountain.
The famous Salmakis fountain in Halikarnassos mentioned by several ancient authors including Vitruvius and Ovid was finally identified by the remarkable "Salmakis Inscription" found by Turkish authorities in 1995. The Danish Halikarnassos project was invited to participate in the study of the remains together with staff of Bodrum Museum and

did so during two visits to the site in 1995 and 1996. The building complex consists of three main structures of Hellenistic and Roman date. In the centre is a well-built Hellenistic wall carrying the Salmakis inscription and in the room south of this is an opening apparently leading into a small cave in the rock. We believe this must be the place of the actual fountain, but it has not yet been excavated. In front of the ashlar wall there must have been a water basin. Mosaics exist both here and in the unexcavated Roman structure south of the Hellenistic fountain house. The Hellenistic building carries the inscription, an elegic epigram of 60 lines, that mentions famous men and mythological matters of importance for the city of Halikarnassos. The inscription was studied and published by S. Isager in 1999. K. Jeppesen, P. Pedersen and B. Poulsen made a brief study of the architectural remains; B. Berg and R. H. Sørensen made a plaster copy of the inscription and secured the Roman mosaics temporarily on the site.

3. Registration of finds
Work has continuously been carried out to register finds from all investigations in town. Up till 1996 the main focus was on material from the Late Antique *Domus* and the Hellenistic House (under the direction of B. Poulsen). Since 1999 work (under the direction of M. B. Briese) has been carried out on findgroups from the excavation at the Myndos Gate, the investigations of the city-wall, the excavations in the castle of St. Peter and the excavations at the rock-cut tombs.

The bulk of material consists of ceramic finds, which are all counted for statistical purposes. Furthermore, all diagnostic pieces are subject to a detailed description. One of the most important aims is to create a typology of the different local waretypes. A fabric most probable to derive from Halikarnassos itself has been defined. This fabric – in varying degrees of coarseness – has been detected in tile, amphorae, cooking ware, domestic ware and fine ware.

4. Sepulchral architecture
A.M. Carstens has conducted fieldwork in connection with her Ph.D. dissertation on tombs in the Halikarnassos peninsula. In 1996 a general survey of especially rock-cut tombs was conducted in the Halikarnassos peninsula in collaboration with the Bodrum Museum. The fieldwork of 1997 concentrated on detailed studies, including measuring of

four chamber tombs. The project has been carried out as a diachronic study of tombs from the Late Bronze Age to late Roman times, and has contributed to the study of cultural interactions and ethnic differentiation characteristic of the Halikarnassos region and south-western Karia. The dissertation was submitted in May 1999.

5. Halikarnassian Pottery
M. B. Briese has worked with finds registration in the project since 1992. Her Ph.D.-project is based upon material from several of the investigated contexts.

Firstly, the aim has been to create a typology of Halikarnassian pottery. In this process Kaare Lund Rasmussen and Karen Berg, both working at the National Museum in Denmark, have contributed with most important results deriving from different scientific analysis. Especially the new method of SUS developed by K L. Rasmussen has shown itself to be most fruitful.

Secondly, total counts from all contexts have been carried out for statistical purposes.

The abovementioned studies together provided a base for a study with the aim to illuminate the economic situation and social life of the city from its foundation until Late Antiquity.

Other works:
B. Berg and his team of conservators have finished the restoration of the Hellenistic mosaics deriving from the Hellenistic House, which was rescue excavated in 1991 (*ActaHyp* 4, 1992, 379). The work of his team has also included the assembling and conservation of a Late Classical grave stele found by construction work below the theatre. Furthermore, B. Berg has been engaged in the conservation of mosaic floors found in the building complexes of the Salmakis fountain.

Bibliography:
J. Isager (ed.), *Halicarnassian Studies I: Hekatomnid Caria and the Ionian Renaissance*, Odense 1994; S. Isager, Pagans in late Roman Halicarnassos 2. The voice of the inscriptions, *Proceedings of the Danish Institute at Athens* 1, 1995, 209-219; B. Poulsen, Pagans in late Roman Halicarnassos 1. The interpretation of a recently excavated building, *Proceedings of the Danish Institute in Athens* 1, 1995, 193-208; S. Isager

& B. Poulsen (eds.), *Halicarnassian Studies II: Patrons and Pavements in Late Antiquity*, Odense 1997; P. Pedersen, Investigations and research in Halikarnassos 1995, *XIV. Arastirma Sonuclari Toplantisi*, Ankara 1997, 207-213; A. M. Carstens, *Death Matters. Funerary Architecture on the Halikarnassos Peninsula*, Ph.d.-dissertation, Department of Archaeology and Ethnology, Faculty of the Humanities, University of Copenhagen 1999; A. M. Carstens, Sepulchral architecture on the Halikarnassos Peninsula, in R. F. Doctor & E. M. Moorman (eds.), *Proceedings of the XVth International Congress of Classical Archaeology* (Amsterdam 1998), *Amsterdam 1999;* S. Isager, The Pride of Halikarnassos, *Zeitschrift für Papyrologie und Epigraphie* 123, 1999, 1-23; P. Pedersen, Investigations in Halikarnassos 1997, *XVI. Arastirma Sonuclari Toplantisi*, Ankara 1999, 325-344; K. Jeppesen, Halikarnass, in: *Der Neue Pauly*, bd. 14, 333-349; K. Jeppesen, *The Maussolleion at Halikarnassos IV: The Quadrangle*, Århus 2000; P. Pedersen, Investigations and excavations in Halikarnassos in 1998, in *XXI. Kazi Sonuclari Toplantisi*, Ankara 2000, 305-314; A. M. Carstens, Drinking vessels in tombs – a cultic connection? in C. Scheffer (ed.), *Ceramics in Contexts, Proceedings of the Internordic colloquium on ancient pottery held at Stockholm, 13-15 June 1997* (Stockholm Studies in Classical Archaeology). Stockholm 2001; P. Pedersen, Report of the Turkish-Danish Investigations at ancient Halikarnassos (Bodrum) in 1999, *XXII. Kazi Sonuclari Toplantisi*, Ankara 2001, 287-298; M. B. Briese & L.E.Vaag (eds), *Halicarnassian Studies III: Traderelations in the Eastern Mediterranean: The Ceramic Evidence*, forthcoming; S. Isager & P. Pedersen (eds.), *Halicarnassian Studies IV: The Salmakis Inscription and Hellenistic Halikarnassos*, forthcoming.

BLACK SEA AREA
43. Excavations at Panskoye, Crimea
Cf. *ActaHyp* 6, 1995, 312-16
Bibliography: L. Hannestad & V.Stolba, *Archaeological Investigations in Northwestern Crimea Panskoye I vol 1, the Monumental Building U6 (forthcomming)*.

ITALY
45. Nemi, loc. S. Maria
(Pia Guldager Bilde)
Direction: scientific committee consisting of the directors of the Nordic Institutes in Rome, Gunver Skytte (Denmark), Christian Krötzl (Finland), Rasmus Brandt (Norway), Anne Marie Leander Touati, from the fall of 2001: Barbro Frizell (Sweden), Giuseppina Ghini (Soprintendenza archeologica per il Lazio), Jan Zahle (Royal Cast Collection, Copenhagen), Pia Guldager Bilde (University of Aarhus, Denmark, field director). Birte Poulsen (University of Copenhagen is director of the find registration).
Financial support: The Carlsberg Foundation (Denmark), the Joint Committee of the Nordic Research Councils for the Humanities (NOS-H) with contributions from foremost the Rausing Family Foundation.
Goal: To excavate and measure a large late Republican to early Imperial period villa situated by Lake Nemi in order to obtain an idea of its layout and architecture, its phases and dating, and its embellishment. Moreover, it is also the purpose of the excavation to evaluate the villa in its local context. This local context comprises the famous Sanctuary of Diana Nemorensis, situated by the northern shore of the small lake, the ancient emissary draining the lake, and the two floating palaces, once belonging to the emperor Caligula, raised from the bottom of the lake between 1928 and 1932. Along with the villa, these elements constitute a most interesting early Imperial period ensemble. The five year-project was launched in 1998.
Results: The S. Maria site by the lake was frequented by Man at least as early as in the Middle (Appenine) Bronze Age as attested by stray finds. Culture layers probably from a settlement dating to the Final Bronze or Early Iron Age have been observed in the northern part of the later villa site. There are no finds from the early Historical periods.

The villa itself had several phases. Its original core was an artificial platform, partly consisting of levelled rock, partly constructed in *opus quasi reticulatum*. The oblong terrace measuring ca. 260 m x ca. 60 m was probably built during the 1st BC. Here the living quartes were to be found, and the basic layout of the buildings on this terrace seems to have been established already in the villa's first phase. The main representative rooms were probably situated in a block in the central zone of the terrace. The northern half of the plateau was articulated with a huge

Π-shaped portico open towards the lake. Behind this portico were series of small *cubicula* (**Plate 22**). Between the main block of rooms and the large portico was apparently the atrium separated from the portico by a narrow block of rooms several of which were furnished with *opus sectile* floors. At least part of the terrace front was seemingly furnished with a columned facade. A monumental Doric capital cut in *peperino* of late Republican style provides us with an impression of its character.

In several trenches we have been able to observe that the walls on the main plateau belonging to the first phase were torn down and only later, in the Imperial period, either reconstructed or overbuilt. This may reflect the strange story told by Suetonius on Caesar constructing an expensive villa in the sacred wood of Diana by Lake Nemi and tearing it down again upon finishing it (Suetonius *Iul.* 46).

During the early Imperial period, the villa was rebuilt and enlarged several times, first between ca. 10-40 AD, and later in the second half of the 1st century AD. In the first Imperial period, the access to the villa was improved with a basalt paved road constructed on a rockcut terrace faced with a mortar wall. A closed peristyle substituting the atrium of the first villa was constructed (**Plate 23**), and a series of *cubicula* were added to the northern end of the villa and the small rooms were furnished with black and white mosaics with geometric patterns.

The second half of the 1st century marked significant building activities in the villa. At the latest during this period a large cistern was constructed at the villa's highest point and probably in its axis. It was built in *opus caementicium* with two barrel-vaulted aisles. The exterior length is 37 m, whereas the internal length is 34 m and the internal width is almost 8 m and its height almost 6 m. Internally and externally the cistern was furnished with nine pilasters securing the stability of the walls. The water was let out from the cistern in (no longer preserved) lead pipes. These pipes were originally conducted in a vaulted mortar channel, no less than 1.2 m wide and 1.8 m high.

In the northern short end and at a lower level than the main villa plateau, a building was constructed against a terrace wall substructing another paved road leading from the villa terrace down to the building situated at a level close to the Nemi lake. This building was at the latest in the fourth phase a bath. Its function in phase 3 is not clear. At the latest in this period too, a long structure consisting of 59 parallel open *fornices* was built against the whole facade of the main terrace wall. Its function was to stabilize the old terrace wall and to carry a carefully constructed channel in which water from the roof of the portico on the edge of the terrace was led away.

Against the mountain in either short end of the villa the building was further monumentalised probably in the second half of the 1st century AD. Towards north, a fornix structure was built against the road terrace, and towards south, an impressive exedra was erected. Both were constructed in *opus mixtum* resembling the technique employed in the later bath building. The exedra was founded directly against the loose pozzolana rock. The two-storeyed, horseshoe-shaped room has a diameter of 21 m and a depth of 17.5 m flanked at the two sides by two two-storeyed wings. The overall width amounts to 48 m, and the visible height of the building is 10.5 m. Due to considerable accumulation of soil inside the room, it has not been possible to reach the original floor level that is probably situated 4-5 m further down. The function and precise date of the room remain yet unknown.

The northern building was at the latest in phase 4 turned into a bath. The lower part of the bath with its service quarters is well preserved, whereas the upper storey has been much damaged through time. However, it may be noted that this part of the bath was once furnished luxuriously with *opus sectile* floors. Tile stamps also *in situ* date the significant building activity to the Hadrianic period during the 120s AD. In this phase the closed peristyle was, furthermore, renovated and furnished with an *opus sectile* pavement consisting of rectangular slabs of Portasanta marble with bands of Pavonazzetto marble, a type known e.g. from Domus Aurea and from Hadrian's villa by Tivoli. This is also true for the larger rooms opening into the peristyle.

Also further rooms were renovated in this period. Very interesting is a *cubiculum* in the western portico. In this room, the original black and white mosaic with geometric pattern was supplied with a new central medallion in the same technique. The medallion featured stylised palmettes and birds, and it was bordered by a band with an inscription stating: M PA(I)MENTVM FECIT, "M(arcus?) made the floor". The anonymity of the mosaicist is puzzling.

Apart from the already mentioned marble and mosaic floors, the villa was once richly furnished. Architectural terracottas (Campana plaques and palmette antefixes), fragments of stucco and wall paintings, thousands of glass and marble mosaic *tesserae*, and revetment plaques for floors and walls crafted in marble imported from all over the Mediterranean attest to its original embellishment. However, the villa has been subjected to severe plundering. This may have started already in antiquity following the abandonment of the villa in the middle of the 2nd century AD.

In late Antiquity, perhaps in the 6th century, the villa site was used for a widespread necropolis with very well constructed inhumation graves reusing building materials from the villa. There have been found no traces of habitation from that period in the villa site. Later, in the Middle Ages, a room in the bath was used as a lime kiln, and abundant marble pieces cut for being burnt to lime have been found in the bath area, including a corner of a relief *pinax*.

Bibliography: P. Guldager Bilde, »Ved Dianas hellige sø. Nordiske udgravninger i Nemi, loc. Santa Maria«, *Hvad fandt vi? En gravedagbog fra Institut for Klassisk Arkæologi*, Århus 1999, 181-194; P. Guldager Bilde, »Cæsars villa? Nordiske udgravninger ved Nemisøen«, *Sfinx* 22.4, 1999, 158-164; S. Fredslund Andersen & P. Guldager Bilde, »Nemi, loc. Santa Maria and the application of computer technologies in field excavation«, *CSA Newsletter* 13.1 2000, http://csanet.org/newsletter/spring00/nls0001.html

Excavation in the Casa di Successo (I.9.3), Pompeii, Italy 1999-2000
(Christina Trier)
Direction: Christina Trier
Financial support: Engineer Sven G. Fiedler and Spouse Scholarship for the promotion of botanical and archaeological research.
Goal: The excavation formed part of a complete analysis of the building development of the four northern houses of insula I.9, in Pompeii. The project was undertaken by C. Trier in 1997, in collaboration with the British School at Rome's documentation project of insula I.9, directed by A. Wallace-Hadrill.
Result: Excavation was carried out in three rooms (labelled 10-12) at the rear of the Casa di Successo. This area already had been partly excavated, but no reports from the previous excavations are available. The aim of the campaign was to establish the floor level and the function of the rooms at the time of the eruption of Mount Vesuvius in AD 79. A further goal was to investigate the development of the rooms and their relationship to the surrounding houses (I.9.8 and I.9.15).

Insula I.9, Pompeji

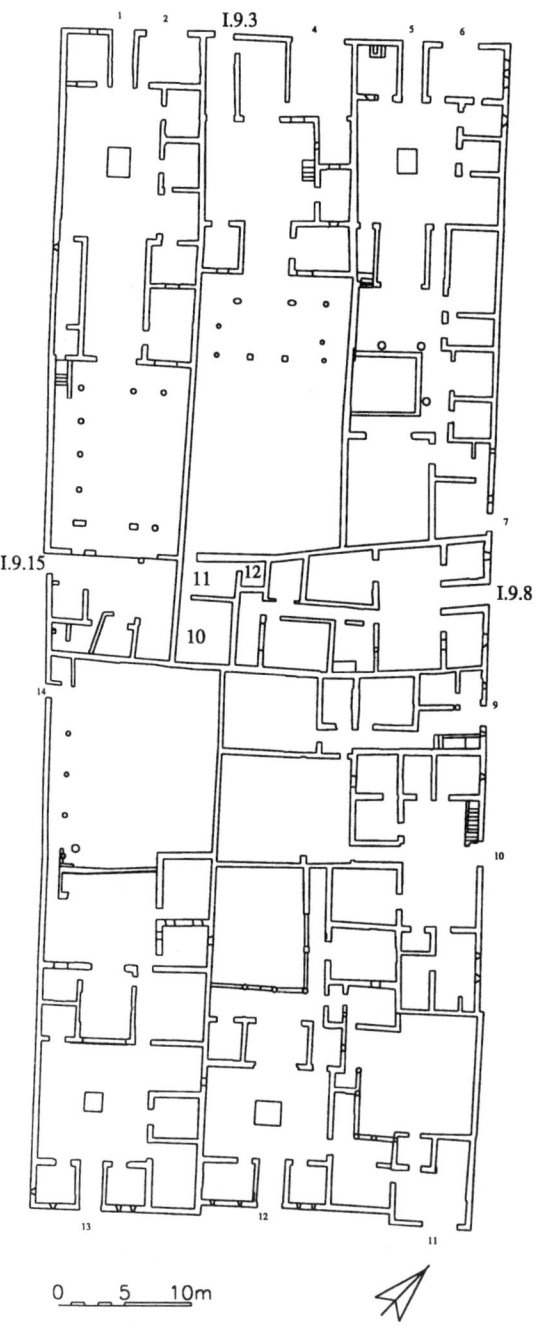

It became clear that all three rooms been excavated in modern times down to the AD 79 level and subsequently covered over again. In room 12 (1.78 x 1.75 m) an intact toilet structure and drain were revealed. The drain was filled with lapilli, indicating that the toilet was in use in AD 79.

In the adjoining room 11 (4.10 x 2.50 m), a floor contemporary with the toilet was found, though it had never received a proper surface. Evidently, construction work was taking place in this room at the time of the eruption. An L-shaped trench (1.00 m wide and 0.60 m deep) conducted along the east and south walls revealed two earlier floor surfaces, the earliest one being made of *opus signinum*.

The floor level of AD 79 was also established in room 10 (4.10 x 4.55 m). Since the campaign season was restricted to four weeks, only limited excavation took place below this surface. Nevertheless, several floor levels and rubble structures were exposed, of which the earliest floor surface corresponded with the signinum floor found in room 11. Unfortunately, a large circular cut in the middle of the room largely obscured the relationship between the floor levels in room 10. Further fieldwork is needed to fully understand the nature of this cut.

The excavation demonstrated that the three rooms had at least four major building phases. Moreover, it confirmed that the rooms were originally part of the adjacent houses I.9.8 and I.9.15, and only subsequently became the property of the Casa di Successo. The transformation took place in two stages, beginning with the blocking of the connection to house I.9.15. The subsequent separation from house I.9.8 and the annexation to the Casa di Successo was related to the last raising of the floor level in room 11, and the construction of the toilet in room 12. Numismatic evidence established a terminus post quem for this rebuilding of AD 36-38, but the strong evidence of construction work in room 11, as well as room10, suggests that the rebuilding took place shortly before the eruption.

The campaign 2000 was devoted to finds study.

Bibliography for the excavations conducted in the southern part of the insula:
M. Fulford and A. Wallace-Hadrill, The House of Amarantus at Pompeii (I, 9, 11-12): An Interim Report on Survey and Excavations in 1995-96. *Rivista di Studi Pompeiani* VII, 1995-96, 77-113; M. Fulford

and A. Wallace-Hadrill, Towards a History of Pre-roman Pompeii: Excavations beneath the House of Amarantus (I.9.11-12), 1995-8. *PBSR* vol. 67, 1999, 37-144.

BOOK REVIEWS

Fernando Rebecchi (ed.), *Spina e il Delta Padano*
Atti del Convegno Internazionale di Studi 'Spina: due civiltà a confronto', Ferrara, Aula Magna dell'Università, 21 gennaio 1994
(Studia Archaeologica 90). L'Erma di Bretschneider. Roma 1998. 358 pages, 150 b/w figs., 12 colour figs. Lit. 380.000.

In January 1994 the international meeting *Spina: due civiltà a confronto*, was held on the occasion of the exhibition *Spina. Storia di una città tra Greci ed Etruschi* shown in the Castello Estense, Ferrara, from 26th September 1993 to 15th May 1994. The papers from the meeting have been published in the present volume under review.

For scholars studying the Po Valley, Etruscan culture, necropoleis, Attic figured pottery, trade and cultural interaction, this publication, together with the exhibition catalogue (here *Catalogue* 1993), will provide interesting and informative reading. It would be too ambitious to give a thorough discussion of all contributions and the following will only highlight the main issues.

Part II presents mythological, historical and archaeological studies of Spina as well as Spina's relation with the Po Valley. The reader might have a better overview if these contributions had been arranged in a thematic (as below) or chronological order, rather than alphabetically according to authors. The first contributions dealing with the mythological traditions on the origins of Spina are by A. Mastrocinque, 'Appunti sulla storia di Spina', (77-84) and F. Prayon, 'Daedalo e Spina', (101-105).[1] Mastrocinque presents the three foundation myths based on written sources: The arrival of the Pelasgians, the Daedalos myth and the Diomedes myth. Prayon, in his turn, views the Daedalos myth from an archaeological angle, tracing two iconographic traditions from the 7th to the 5th century BC, a south Italian and a local variant where Daedalos is characterized by his double function as a hero and a constructor/architect.

The political and economic roles of Spina are treated by D. Briquel, 'Aspetti politici', (41-52), and M. Gras, 'Spina: Aspetti commerciali', (57-64).[2] Briquel deals with the much debated topic regarding the dependence (or not) of Spina to Bologna and the range of power of the

Greek community in Spina. The author argues that Spina should be seen as a multi-ethnic society (Etruscan, Venetic, Umbrian and Greek), having a specific mercantile role in the *dodecapolis* federation of the Etruscan Po Valley. Based on parallels with Mantua, Briquel tentatively suggests that the Greek *ethnos* would have had some political influence (49-50). Gras discusses the function of Spina with the frequently asked question: town or *emporion*? Initially under the influence of Bologna, Gras believes that Spina became a harbour with a certain autonomy and with no Etruscan nor Greek parallels. He suggests that Spina became an independent harbour that might be compared to Ostia, an interesting aspect which will be constantly debated as no single term can define Spina. Its function should rather be seen as varying through its c. 200 years of existence. Spina was the harbour of the Po Valley and dependent on trade. Periods with much trading activity and economic independence were possibly followed by political independence.

The necropolis is only discussed and analyzed by B. d'Agostino, 'L'immagine della città attraverso le necropoli', (53-56).[3] He mainly deals with the necropoleis of Spina and Bologna and quickly concludes that since the necropolis (of Spina) was formed by plots, this reflects a stable community with no competition on the social scene; a rather sketchy study which does not take the complex time- and space pattern of the necropolis into consideration.

The immense quantity of pottery finds has surprisingly only encouraged two scholars to present some stimulating thoughts on both the imported and local production: F. Lissarrague, 'Spina: Aspects iconographiques', (67-75) and J.-P. Morel, 'Su alcuni aspetti ceramologici di Spina', (85-99).[4] Lissarrague attempts to ascertain iconographic patterns on Attic figured pottery found in a selection of graves from Spina. Although a difficult task, some important points are put forward: firstly the difficulty to analyze iconographic patterns due to the large dating discrepancies in some graves; and secondly, that the trade appears to have been directed by shapes rather than by motifs (72), although without mentioning which shapes were popular. A most inspiring point is the possibility to follow the pottery of various workshops in time and space. Twenty vases by the Half Palmette Painter were found in 16 graves. Some had been bought and placed immediately in the grave, others passed a generation or so above ground, some were bought by wealthy individuals (judging by the grave goods), others made part of modest grave offerings (72-74). Morel works with a somewhat neglected part of the ceramic corpus; the local pottery and the imported black-glazed pottery. The local pottery appears to be more influenced by Etruscan pottery than by Greek imports and there was a predilection for metallic prototypes, especially the phiale. Concerning the imported ware the shapes found in Spina are very different from finds in North-African and Spanish sites. As with the Attic figured pottery this may show a specific demand from the customers in Spina (93).

The last contribution is on the Roman Po Valley by S. Rinaldi Tufi, 'La decadenza e l'età romana', (107-110).[5] The article deals with the role of Ravenna after the decline of Spina and the importance of the *fossae* further developed by the Romans in the marshes of the Po Valley.

Part III consists of twelve presentations and comments made at the meeting discussing the above papers. The main arguments are the significance of the Daedalos myth, the inscriptions found in Spina and their interpretation, and the political-economic status of Spina.

The discussion on the Daedalos myth is complex and this is not the place to treat it in depth. Briefly, it can be said that according to Pseudo-Aristotle (836 A-B-*De Mirab. Ausc.* 81) Daedalos reached the northern Adriatic after his flight from Crete. This has been connected with various archaeological finds from the Po Valley figuring Daedalos. The role of the Daedalos myth in the Po Valley is thus frequently stressed, but not agreed upon. Colonna (127-130), who has worked with the *graffiti* from Spina since 1977, stresses that any statistical calculations on the corpus is futile as all have not been published. In spite of this, Colonna stated that 27 % of the inscriptions are Greek. It is important to keep in mind that the number is based on graffiti from vessels dated from 500 to 350 BC, found in 13 graves as well as in two places in the settlement, and that graffiti does not *per se* equal ethnicity. To achieve a wider picture of the graffiti it is necessary to define their chronological distribution, to compare them with the total amount of Etruscan graffiti and to define whether they are owners' names, dedications to gods, or whether they served other purposes.

Sassatelli (160-164) focuses on the discussions regarding the status of Spina and finds most of the theories unsatisfactory, as they indicate either a dependence on the *hinterland* or a complete independency as an autonomous town. Sassatelli argues that the complex structure of the Po

Valley has not been considered. The Po Valley differed from Etruria proper, as all the towns and settlements were complementary to each other and as a rivalry between towns as seen in Etruria did not exist. The towns of the Po Valley were connected through the ties of trade stabilized in an economic and political federation, and one town could not exist without the other. As such Spina was as (in)dependent as any town.

Part IV contains articles on archaeological topics and short notes on recent excavations in the Po Valley. The first article is by M. P. Guermandi, 'Figure in quantità. L'analisi quantitativa della ceramica attica in Spina' (179-202). Her analysis is based on a sample of 2200 figured vases, which includes all vases listed by Beazley, *CVA*, and the main publications of Spina. It is estimated that as many as 10-12 000 vases have been found and that far from all of these have been published. From the sample it was possible to ascertain the chronological distribution of shapes and painters. As one of the few pottery quantification studies she also considers the relations between motifs, shapes and painters. It appears that shapes rather than motifs were traded, that the trade was not affected by the Peloponnesean war, and that motifs were related to shapes: e.g. athletes predominantly occur on kylikes and oinochoai, gynecaeum-scenes on oinochoai, kylikes and lekanai, and ephebes on kraters. Research on such an impressive material is invaluable. Yet, while quantification studies may show general patterns in the material, it is important to stress that they do not show why such choices were made, how the pottery was used, whether some graves contained figured pottery assembled with a specific figural theme, comparisons with local pottery or parallels with other sites. Nor do they take the context of the pottery into account when indeed Spina is unique in having c. 4000 graves with known contexts.

The aspect of pottery linked with trade is dealt with by S. Bruni, 'Un problematico documento per la storia della frequentazione dell'area spinetica prima di Spina. Appunti sulle rotte adriatiche in età arcaica' (203-220). Bruni investigates the circulation of Corinthian pottery in the Adriatic, which he links to the finds of Laconian pottery. However, the distribution maps (figs. 3-4) of Corinthian and Laconian finds in the Adriatic appear to be direct opposites, despite Bruni's argument that the "distribuzione dei materiali laconici....sembra ricalcare gli stessi circuiti segnalati....per i materiali corinzi" (212-213).

No sanctuaries or votive deposits in Spina have yet been excavated, but Colonna, 'Il santuario extra-urbano di Spina in Loc. Cavallara' (221-226), attempts an identification of a sanctuary based on dispersed finds of bronzes supposedly found in the Loc. Cavallara. If the identification is correct, the site can be interpreted as an extra-urban sanctuary of Spina worshipping Hercules and possibly also other gods.

A study of the Etruscan graffiti is presented by A. Maggiani, 'Sulla paleografia delle iscrizioni di Spina'(227-234). Altogether 180 Etruscan graffiti have been found in Spina, less than 130 of which are published. Maggiani traces the origin of the letter forms, their diffusion in the Po Valley and the earliest examples in Spina.

Three recent excavations in Polesine, Adria and Case Nuove di Siccimonte are presented by R. Peretto – L. Salzani,'Polesine: le recenti scoperte' (235-240); S. Bonomi, 'Adria e Spina' (241-246); M. C. dall'Aglio,' L'insediamento etrusco di Case Nuove di Siccimonte (Parma)' (247-251). The site of Balone in Polesine was excavated in 1987-90. Four inhumation graves dated to the fifth century BC were found as well as a part of the settlement. The graves contained red-figured pottery, *etrusco-padano* ware and Etruscan bronzes, while in the settlement (probably huts) only plain ware was found. Trade routes connected the site with Adria and the *hinterland*. Adria has until recently only been archaeologically known through the fourth century BC necropolis. The town, having been inhabited continuously up to this day, has not been excavated in its entirety. Due to an emergency excavation, 20 graves (both inhumation and cremation) dating from the end of the sixth to the end of the third century BC were found. These, although few, will enable scholars to make comparative studies between Adria and Spina, and perhaps also define the relation between the two coastal towns. The recently excavated Etruscan site of Case Nuove di Siccimonte may also prove interesting. Traces from five huts, clay-pits, small canals and one well were found. The huts were made in different plans. The oldest, dating to the sixth century BC, was square and can be compared to the Etruscan house found at Forcello near Mantua. An abundance of pottery was found, predominantly local coarse and fine wares, but also bucchero (local?) and one sherd inscribed with the Etruscan letter U. Also fragments of Attic figured pottery were found, all dated from the sixth to the beginning of the fourth century BC. The site (together with the abovementioned Balone) provides interesting information on Etruscan settlements. This is an important new aspect of the Etruscan culture, which is otherwise best documented by the necropoleis. Especially the finds of pottery can refine existing typologies of the local production, and also testify how the Attic pottery was used in the household compared to what we find in the graves. Further excavations in the Po Valley will surely provide more evidence of this kind. Case Nuove di Siccimonte (near Fidenza) also confirms the expansion of Etruscan culture towards the north-western Po Valley, which little archaeological evidence has shown so far.

The following two articles, D. Vitali, 'I Celti a Spina', (253-273) and N. Camerin, 'I celti anche nel Delta padano?', (275-283), deal with related topics. The 'Celtification' of the Po Valley is mainly based on a passage by Dionysios of Halicarnassos (1, 18, 5) describing the invasion of the 'Barbarians' around 388/384 BC. The archaeological evidence of Celtic presence in Spina, though, is very scarce. Four graffiti on pottery, presumably with Celtic names, dated to the second half of the fourth century BC have been found. No graves can be interpreted as typically Celtic, nor can they be compared to any of the graves found in the Celtic site of Monte Bibele (fourth to second century BC). Any indication of a Celtic invasion should therefore be treated with caution and it is likely that the two peoples shared common commercial interests and were able to live in peace with each other.

An attempt to estimate the population of Spina is made by D. Pupillo, 'Spunti per una ricerca demografica su Spina' (285-293). This is based on the (estimated) size of the site and the number of excavated graves. An interesting point, if it is indeed possible to ascertain the human resources of Spina and the impact its population may have had on the rest of the Po Valley. However, the calculations are based on very scarce evidence (as stated by Pupillo herself) and, although such calculations are not futile, they should be treated with great caution due to their hypothetical character. Estimations based on the site vary from 1665 to 3528 individuals (for what period?), whilst the necropolis has yielded c. 4063 individual graves dated c. 500-250 BC. Pupillo may be right in stating that the population varied seasonally depending on harvest and trading periods, but this cannot be documented by the archaeological material. Periods with flourishing trade may also have affected the demography. This can, in my view, be seen for the period 425-400 BC, as this period experiences a rise in the number of burials and an increasing prosperity, judging by the grave goods. To divide the burials and the (estimated) population into chronological periods would be useful, since it is possible to compare periods of rise and decline with contemporary historical events and to establish parallels to other sites.

Finally the Roman period is discussed by F. Rebecchi,' "Grecità" e Greci a Ravenna (e dintorni): novità ed elementi di discussione' (295-321). Rebecchi deals with Roman Ravenna at a time when Spina had ceased to exist. Signs of a hellenization and of Greek residents can also be seen here. The Greek presence is evidenced by a late funerary monument (second century AD) preserved with inscriptions of a certain T(ito) Ioulios N(e)ikostratos from Rhodes.

Part V contains conclusive remarks by P. G. Guzzo and P. E. Arias given at the meeting, which would have been more suitably placed directly after part III. Finally, there are indexes compiled by M. Harari and F. Rebecchi of ancient authors, inscriptions mentioned in the texts, modern authors and index of persons and places.

Parts II-III and conclusions in part V clearly use the *Catalogue* 1993 and the exhibition as their frame of reference. This must have been very productive in a forum of scholars who had the opportunity to discuss various points of views (all the contributors of the catalogue were present), but for the reader (who was not there) the most innovating talks appear to be those discussing aspects that were only touched upon in the catalogue. The intention of part IV is to present recent studies and excavations in Po Valley, although the excavations of Balone and Adria have by now been partly published.

Nevertheless, the publication has been worth waiting for and covers most aspects relevant to Spina. Contributions in part II and III, however, are very much linked to the Catalogue, which is indispensable when reading the present volume under review.

As the title of the meeting suggests "Spina: due civiltà a confronto", it was intended to discuss the relationship between the Etruscan and the Greek culture at Spina. This aspect, however, appears to have given way to more regional studies regarding Spina and the Po Delta. It would have been very interesting to know what classical 'Greek' archaeologists would theorize about the burial-rites, the function of imported pottery, the graffiti, the question of *emporion/apoikia* compared to, for example, Greek colonies etc. In this respect the meeting deceived its title. The contributions, 31 in total, including papers and discussions, are mainly by Italian archaeologists with five French exceptions and one German, and except for two contributions in French, all are written in Italian. Also the list of participants (18) underlines that dealing with Spina is still widely limited to the Italian circle of archaeologists. It is to be hoped that the publication will open up for a broader international discussion.

NOTES

1. Cf. M. Torelli, Spina e la sua storia, in *Catalogue* 1993, 53-69; L. Braccesi & A. Coppola, I Greci descrivono Spina, in *ibid*, 71-79; G. Sassatelli, Spina nelle immagini etrusche, in *ibid.* 115-127.

2. Cf. G. Colonna, La società spinetica e gli altri ethne, in *Catalogue* 1993 131-143; L. Malnati, Le istituzioni politiche e religiose a Spina e nell'Etruria padana, in *ibid.*145-177; G. Sassatelli, La funzione economica e produttiva: merci, scambi, artigianato, in *ibid.*179-217.

3. Cf. P. G. Guzzo, Ipotesi di lavoro per un'analisi dell'ideologia funeraria, in *Catalogue* 1993, 219-229.

4. Cf. P. G. Guzzo, Vasi attici a figure, anche a Spina, in *Catalogue* 1993, 81-213.

5. Cf. F. Rebecchi, Il delta adriatico in età romana, in *Catalogue* 1993, 233-245.

Alexandra Nilsson
Fredericiagade 79 st. th.
DK-1310 Copenhagen K

Tarquinia. Scavi sistematici nell'abitato. Campagna 1982-1988 I materiali 1. A cura di Cristina Chiaramonte Treré. (Università degli studi di Milano. Tarchna II). Pp. xxxiii + 402, color pls. 3, pls./drawings 86. "L'Erma" di Bretschneider, Roma 1999. Lire 550.000. ISBN 88-8265-068-5.

The plateau, known as the Pian di Civita, north-east of the present-day Tarquinia was the site of the ancient Etruscan city of that name. There have been several sporadic excavations of isolated areas; this not yet completed three-volume work constitutes the meticulous report of the six-years-long work on one such isolated area of the city, measuring ca. 24 by 38 metres at the most. The Italian archaeologists refer consistently to it as a *complesso* or an *abitato*. The area must have served a sacred purpose, as the evidence shows, both in pottery and in some splendid bronze objects found in votive deposits from the early seventh century BC.

On the site are several wells as well as remnants of structures. Masses of terracotta sherds have been found: fragments of architectural tiles, of plain and decorated pottery, etc. There is solid evidence that the complesso had a long history beginning in the tenth century and ending as late as in the Hellenistic period. Its heyday seems to have been in the seventh and sixth centuries BC, when it is referred to as having undergone a "monumentalization".

The excavations are of particular importance. The vast majority of Etruscan finds have come from necropoles; evidence from inhabited areas is much more rare.

The first volume (1997) (reviewed in *ActaHyp* 8, 2001, 280-285 by Helle Damgaard Andersen) primarily records the process of the excavations and the interpretations of the historical phases, while the volume here under review (1999) and the projected last volume are devoted to objects found, *i materiali*. Some of the objects were already published in preliminary manner in *Gli Etrusci di Tarquinia* (1985) by some of the same scholars who are involved in the present project, notably Maria Bonghi Jovino and Cristina Chiaramonte Treré; other specialized articles have appeared elsewhere.

Initially Bonghi Jovino explains how the study of the individual groups has been constructed, and then follows a comprehensive descriptive catalogue in the six principal sections: the terracottas (1-41); impasto pottery from the Archaic and Hellenistic periods (43-97); fine unpainted and banded pottery (99-176); Etrusco-Corinthian pottery (17-204); black-glazed, Hellenistic pottery (205-259); and amphoras (261-278). In addition to the illustrations the volume contains 49 pages on technological analyses of the impasto and the black-glazed pottery respectively, followed by 63 pages of indices and tables of correspondencies.

Silvia Ciaghi deals with the numerous, extremely fragmented tiles (the earliest known roof is dated to the first half of the seventh century) divided into two groups, structural and decorative tiles. These are analyzed with respect to both form and fabric with an emphasis on antefixes and frieze plaques. Here she carries out useful comparisons to similar or identical specimens from other sites. One fragment shows a hoplite in the act of getting on a chariot. She recognizes this as a part of a decorated frieze plaque known in its entirety from Aquarossa (type B) and also from fragments in Tuscania. Yet another fragment with meanders and birds has parallels in Rome, Veii, and Velletri. Thus more becomes ascertainable about the distribution or travels of these two matrices. They are thought to have decorated the main building (*edificio Beta*).

The clay used for the tiles has also been analyzed; in all probability it is a local product.

Ciaghi's able and stimulating article is a daring one, for, like her co-authors, she ventures reconstructions and comparisons on the basis only of thoroughly fragmented pieces, as she herself readily points out. Her reconstructions are in fact only possible on the basis of comparisons with finds from other sites.

Cristina Chiaramonte Treré discusses the impasto ware from the Archaic and Hellenistic periods. She begins by defining impasto as all pottery produced in coarse or unpurified clay of local manufacture from the tenth to the second century. Her investigation comprehends 1000 objects from the seventh to the second century BC (an investigation of the impasto ware from the preceding period, by Bonghi Jovino, will appear in the third volume). Impasto in general is produced in vast quantities and repeatedly employs the same shapes throughout the period in question. Thus dating is sometimes difficult and stratigraphical analysis not conclusive: a stratum may have been interfered with in one way or another.

One of the most productive methods of dating is that of comparing findings from this site with those of other sites, where dating is more certain.

Chiaramonte Treré asks four questions of her material: where, how, when, and why the vessels were made. Answers to the first two may be obtained from technical analyses carried out on a corpus of selected fragments (by Ninina Cuomo di Caprio 312-329). The mineral composition of the clay demonstrates that it has been dug in the vicinity of Tarquinia. It becomes more difficult to determine whether a pot was turned on the wheel or not. Here, as throughout the volume, the material is classified morphologically.

Among the closed forms the most common are *ollae* for practical household uses, such as cooking, in storage rooms, or to serve meals; occasionally they were also in use at ceremonies. The most numerous type among the open forms is that of basins – also these occasionally employed in ceremonies.

Chiaramonte Treré should indeed be praised both for her uncompromising honesty and her care. She makes it clear that this type of pottery, the cheapest and the most ubiquitous, is exceedingly difficult to be precise about, and this conclusion must be a comfort to all archaeologists confronted with an odd piece of impasto. That she also manages to single out some special rims in sacral contexts makes her work all the more impressive.

In order to avoid repetition, allow me to state unequivocally that, like the first two authors, the remaining four are equally able, learned, and daring when they construct their hypotheses.

Giovanna Bagnasco Gianni is responsible for the report on the Etruscan unpainted and banded pottery made of pure, fine quality clay found locally in its natural state and on the more numerous, similarly constituted and shaped imported ware from Eastern Greece. The domestic goods imitate the foreign ones; they are undecorated or adorned with painted monochrome bands that are familiar to us from the western Mediterranean area from the end of the seventh century.

Some of the types, e.g. cups and plates, have been adapted from eastern Greek standard shapes; others are derived from the Etrusco-Latial area, and yet there is a certain continuity in the development of the shapes, which may be seen as a consequence of the considerable influence exerted by the imported goods. Here Gravisca, the port of Tarquinia, becomes an important factor, for at Gravisca there is an important amount of imported banded pottery as well as a considerable number of local products imitating them. The author also looks for comparable evidence elsewhere, for example at Caere or Pyrgi. There is much corroborative evidence as far away as at Pontecagnano, in Greece, Spain, and France.

Giuseppina Sansica deals with Etrusco-Corinthian pottery. The material is limited to 100 badly damaged fragments. Open forms, mostly plates, predominate. Local manufacture of this type of pottery was commenced in Tarquinia early in the sixth century, which is to say later than in other Etruscan centres. Still, the majority of the specimens can be shown to have come from Vulci.

Tarquinia does not appear to have hosted major vase painters, and the quality of the work done there was mediocre. This is a striking fact, especially if the brilliant tomb paintings are considered (cf. 195, n. 116). Three separate workshops have been identified, but only one, *"il Gruppo del Pittore senza Graffito"*, is represented at the site. As is the case with the banded pottery, many of the Etrusco-Corinthian plates and basins were in use in the sacral area (vol.I, 194). It might finally be remembered that in spite of the poor quality, this type of pottery was exported, for example to Carthage and southern France.

The following section on the Hellenistic black-glazed pottery is by Magda Niro Giangiulio. Here the findings are more copious; ca. 1500 fragments have been excavated, but their condition is so bad that their shapes only can be ascertained in some 300 cases. The extensive and complex technical analyses (by N. Cuomo di Caprio, M. Pichon, A. Cesana, and M. Terrani, 281-329) aim for morphological and chronological classifications of the material, but should also chart local production of black-glazed pottery, principally from the outset of the third century. In this way it has been possible to show that one or more Tarquinian workshops – the exact locations still unknown – did produce vases with shapes and decorations that are familiar from other sites. In the conclusion Giangiulio attempts to narrow down what might be called the special Tarquinian characteristics of this type of pottery.

The final section on amphoras is by Cecilia Scotti. In all 70 fragments of domestic and imported amphoras from the seventh to the second century have been found. Among the latter, there are amphora fragments of Etruscan, Phoenician/Punic, Attic, Ionic-Marsiglian, Chiotian, Samian, Clazomenian, Graeco-Italic and Corinthian origin. Such an international scope testifies once more to the importance of Tarquinia as a major commercial centre. Gravisca must have been an important harbour. Still, it should be remembered that Cerveteri, for example, has yielded much higher numbers of amphoras – whatever the reason.

It will be clear from the above that this is a most impressive work, a model for years to come. The bibliography (pp. xiii-xxxiii) alone is the most adequate one that this reviewer has encountered. Great care has been lavished on every conceivable aspect of these excavations and that care has been the work of brilliant scholars. There is solid uniformity to the individual contributions; each contributor knows her stuff and is at the same time fully aware of the others' work. They have been blessed by their publisher, for the excellence of the archaeological work is matched by immaculate printing and editing. The format is indeed as appealing as are the contents.

(It would seem picayune to harbour any criticism, but here it comes anyhow. The writers are scholars and do in fact write as scholars should. But the ornate, Italian academic style often comes dangerously close to being turgid, and convoluted; it is as if simplicity were a vice to be shunned at any cost.)

Margit von Mehren
Department of Archaeology and Ethnology
Vandkunsten 5
DK-1467 Copenhagen K

RENATE ROLLE UND KARIN SCHMIEDT (Hrsg.)
in Zusammenarbeit mit Roald F. Docter: *Archäologische Studien in Kontaktzonen der antiken Welt.*
Veröffentlichung der Joachim Jungius-Gesellschaft der Wissenschaften Hamburg, Nr. 87, Göttingen 1998. 886 pp., 66 plates, and text figures.

The 63 articles collected in the volume are dedicated to Prof. Hans Georg Niemeyer of the Hamburg University in honour of his 65th birthday. It forms an impressive collection of studies ranging in time from the second millennium BC to the second millennium after the beginning of the Christian era, and in space from the Eastern Mediterranean to the USA. Although most of the studies can be said in one way or another to deal with the complex issue of contact, the volume is a good example of how many ways this problem can be approached, both thematically and methodologically. The majority of the articles are written in German, but also English, Spanish, Italian and French are used.

It is of course impossible to discuss or even summarise each article from the volume, but as the many Phoenician contributions surely reflects the interests of Prof. Niemeyer and his inspiration of younger scholars, some should be mentioned as examples of the wide range of studies in this compilation.

As can be expected in a compilation of mainly archaeological contributions, the majority of the articles deal with the indestructable pottery, and in this case – hardly surprisingly – studies of pottery from the 8th – 7th century BC dominate. From Greece in the East to the Iberian Peninsula in the West, imports and imitations of Phoenician artefacts (notably pottery) are important for the interpretation of the Iron Age interactions of the Mediterranean and the Black Sea region. With this in mind it is unfortunate that similar studies are lacking from the Levantine coast. Among the "Greek" contributions we find *Neeft*'s purely typological study of a group of Corinthian pottery centred around the Fledgeling painter, and the same method is employed by *Gercke* to identify the cultural sphere of an East Greek bronze thymiaterion. Based on a typological description of Fikellura amphorae found at Miletos and scientific analyses of the material, *Seifert* argues that the group was produced in the Miletos area and probably exported. According to

Morris & Papadopoulos the Phoenicians played a very prominent role in Corinth. Not only was "the Corinthian pottery industry ... determined and defined by the Phoenicians", but the Phoenicians are also believed to have been deeply involved in the establishment of early trade centres in the Aegean (Lefkandi, Knossos, Athens), and their influence in Corinth is seen in, among others, the early sanctuaries and reflected in the foundation myths. Among the Greek contributions we also find *Kourou & Grammatikaki*'s discussion of the issue of Phoenician presence and influence in the Aegean area as tradesmen/navigators *vs.* immigrants/craftsmen, with the discussion taking its starting point in a newly excavated Phoenician stele found in secondary position in the North Cemetery of Knossos.

The Central Mediterranean triangle (Tunisia/Sicily, Sardinia and Tyrrhenian Italy), previously discussed in M. Gras' *Traffiques Tyrrheniens* (1985), is indirectly underlined by, for example, Docter, Durando and Rathje. *Docter* presents distribution charts of the five subgroups of the ZitA amphorae and uses the distribution pattern as a basis for ascription of the various subgroups to different production centres. With this basis he discusses the role of the Phoenicians as mediators in the trade with indigenous peoples of Sardinia and the Italian peninsula. Durando presents results of archaeometrical tests of Phoenician and local amphora from Pithecoussai, and compares each of the ten analysed amphorae with formal parallels from other sites. *Rathje* analyses "a rare decoration on a rare vessel", and apart from pointing out the evident merging of elements from different Mediterranean centres: the Phoenician decoration, the Greek vase form (the *kotyle*), and the Central Italian production of *impasto sottile*, she asks why, tentatively suggesting that the exceptional vessels may have been used in a ritual.

Among the articles dealing with the Far West we find *Mansel*'s study of handmade Iberian pottery found in Carthage. *Ramón* traces Carthage's economic interests in Spain from the 8th century *via* a Carthaginian red-slip plate from c. 675 BC found in Alicante, and similar finds of Carthaginian pottery on the Iberian Peninsula. *Maass-Lindemann* uses the imported Phoenician pottery from the homeland to date the earliest Phoenician settlements in the Iberian Peninsula in the 8th century, and she notes the chronological correspondence with the extremely few finds of sherds of Greek pottery in Spain.

The articles discussing aspects of religious life in antiquity reflect the focus on cult and ritual in contemporary archaeology. Again a wide variety of approaches are employed. Some contributions deal with Phoenician shrines abroad: *Shaw* interprets the shrine at Kommos as an example of Phoenician influence on Crete around 800 BC, and *Karageorghis* presents material to support his identification of the Kition shrine as dedicated to the Phoenician Astarte. *Thomas* presents a study of Greek sanctuaries in the earlier Bronze Age palaces and argues for a continuity of cult on these sites, while *Höcker* traces the genesis of the monumental Ionian temples in an architectonic study.

Another aspect of ritual is the burial. The ceremonies at the tombs are clearly ritual, and the burial ground is a sacred area. In fact, a distinction between the two may be hard to make, as is evidenced by *Briese's* investigation of the interpretation of the pottery from the *Chapelle Cintas* – foundation deposit or burial ground for the first generation of settlers?

The editors have chosen a mainly geographical division of the articles (Orient, Ägypten und Zypern; Griechenland, Ionien und die Ägäis; Griechen und Phönizier im Tyrrhenischen Raum; Karthago und der nordafrikanische Raum; Mittelmeerkonntakte der iberischem Halbinsel; Mitteleuropa und das Schwarzmeergebiet). This to some degree obscures the fact that the vast majority of the articles discuss problems concerning Phoenician/Punic archaeology – the field in which Prof. Niemeyer himself has contributed so much, e.g. in *Die Phönizier und die Mittelmeerwelt in Zeitalter Homers* (1984). More importantly, this division seems to be in contrast with the commonly accepted view that the Phoenicians from the Late Bronze Age onwards were one of the prime mediators in the interrelations criss-crossing the Mediterranean. The chosen presentation also blurs the fact that articles dealing with the "post-Archaic" period do appear in the first sections: *Jacobs* attempts to make a relative chronology of the sanctuaries of Antiochos I; *Meyer* discusses the interpretation of a Hellenistic marble figurine; *Bechtold* provides a preliminary presentation of the excavations in the Hellenistic cemetery at Segesta; and *Lund* argues that a fragment of a Roman historical relief now in the Danish Nation Museum may come from a Flavian monument from the Forum of Carthage.

Still, the geographical division outlined above is not strictly followed. In the second part of the volume there are sections on Die iberische Halbinsel unter den Römern; Rom und seine Provinzen, and the last short section on *Antikenrezeption* has only two contributions. This change from one classification system to another highlights the problem faced by the editors: how to make a homogeneous volume out of a huge number of detached articles. Personally, I would have preferred the contributions to have been presented in a strictly chronological order, to enable the readers to follow the historical development of the Mediterranean from a Phoenician point of view in close connection with the contemporary events in other areas. Admittedly, the inescapable lacunae would have stood out more clearly in this way.

All in all, there is much food for thought in *Festschrift Niemeyer* and the volume stands as a monument to a great scholar.

Helle W. Horsnæs
Ndr. Fasanvej 201, st. tv.
DK-2000 Frederiksberg

kollhors@worldonline.dk

PLATES

Plate 1. Lip and side fragment from an Attic red-figure bell krater. Section with gods in procession. The Altamura Painter. Height 17,2 cm. From Locri Epizephyrii. The Thorvaldsen Museum, Copenhagen.

Plate 2. Neck fragment from an attic red-figure volute krater. Fleeing females at an altar. The Niobid Painter. Height 11,3 cm. From Locri Epizephyrii. The Thorvaldsen Museum, Copenhagen.

Plate 3. Bull protome. Bronze. Height 3,47 cm. Presumably from the Persephone sanctuary in Locri Epizephyrii. The Thorvaldsen Museum, Copenhagen.

Plate 4 a-d. Satyr head. Bronze. Height 2,4 cm. Findplace unknown. The Thorvaldsen Museum, Copenhagen.

Plate 5. Section of an Attic black-figure oinochoe, showing satyrs abducting nymphs. Height of figure panel 11 cm. The Thorvaldsen Museum, Copenhagen.

Plate 6a. CP 2.38.1036. Fragment of medallion with small part of sakkos.
Plate 6b. CP 2.27.1012. Fragment of medallion with part of woman's head.
Plate 6c. CP 2.27.1014. Fragment of medallion with tip of nose and part of rim.
Plate 6d. CP 2.27.1021. Fragment of medallion with part of ear and earring.

Plate 7a & b. CP 2.42.1004. Fragment of medallion with part of woman's face and base with graffiti on the lower surface. Drawing: the author.

Plate 8a. CP 2.27.1013. Fragment of medallion with dotted rosette and part of ray.

Plate 8b. CP 2.37.1017. Fragment of medallion with part of star with striped chevrons between the rays.

Plate 8c. CR 5.1.749. Rim fragment with cross-hatched decoration.

Plate 8d. CR 5.8.632. Rim fragment with laurel wreath.

Plate 8e. CF 2bis.371. Rim fragment with laurel wreath.

Plate 9. House model from Basilicata. The Danish National Museum, Dept. of Near Eastern and Classical Antiquities, inv. no. 3732 (courtesy of the Dept. of Near Eastern and Classical Antiquities, photo by Lennart Larsen).

Plate 10. Dinos, Staatliche Kunstsammlung Kassel, inv.no. Alg. 51 (from Gercke 1981, 71).

Plate 11. Cypriot transport amphora with horizontal handles from Marion, tomb 80 (Department of Antiquities inv.no. 9706, National Museum, Copenhagen).

Plate 12. Red Slip IV (VI) ware bowls from Marion, dated between ca. 480 and 400 BC. The shape of the example to the right - notably its hooked rim - has an uncanny resemblance to that of Cypriot Sigillata form P 22 A.

Plate 13. The phiale from J.F. Willumsen's Collection, after the reassembling at the National Museum of Denmark. H. 2.8 cm, diam. 9.0 cm (Photo by Bent Ryberg, courtesy of the J.F. Willumsen Museum).

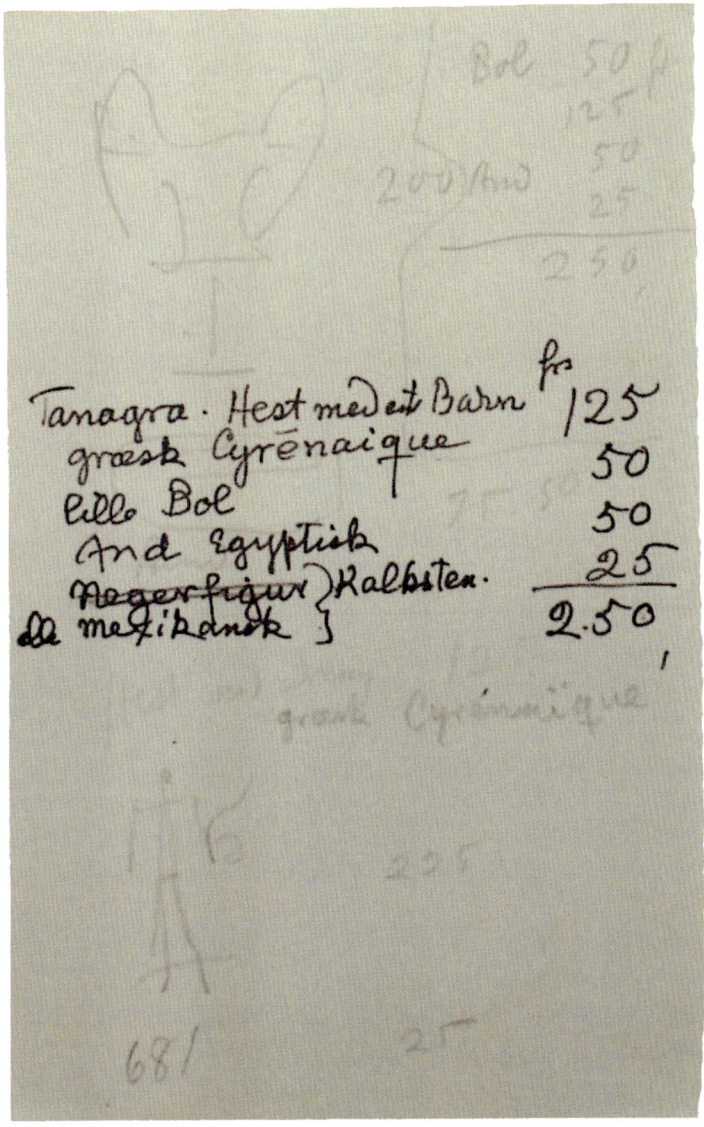

Plate 14. Note written by Willumsen recording prices for the "Small Bowl" and other pieces of art (photo by Kaare Lund Rasmussen).

Plate 15. Vroulia (courtesy of the Dept. of Near Eastern and Classical Antiquities).

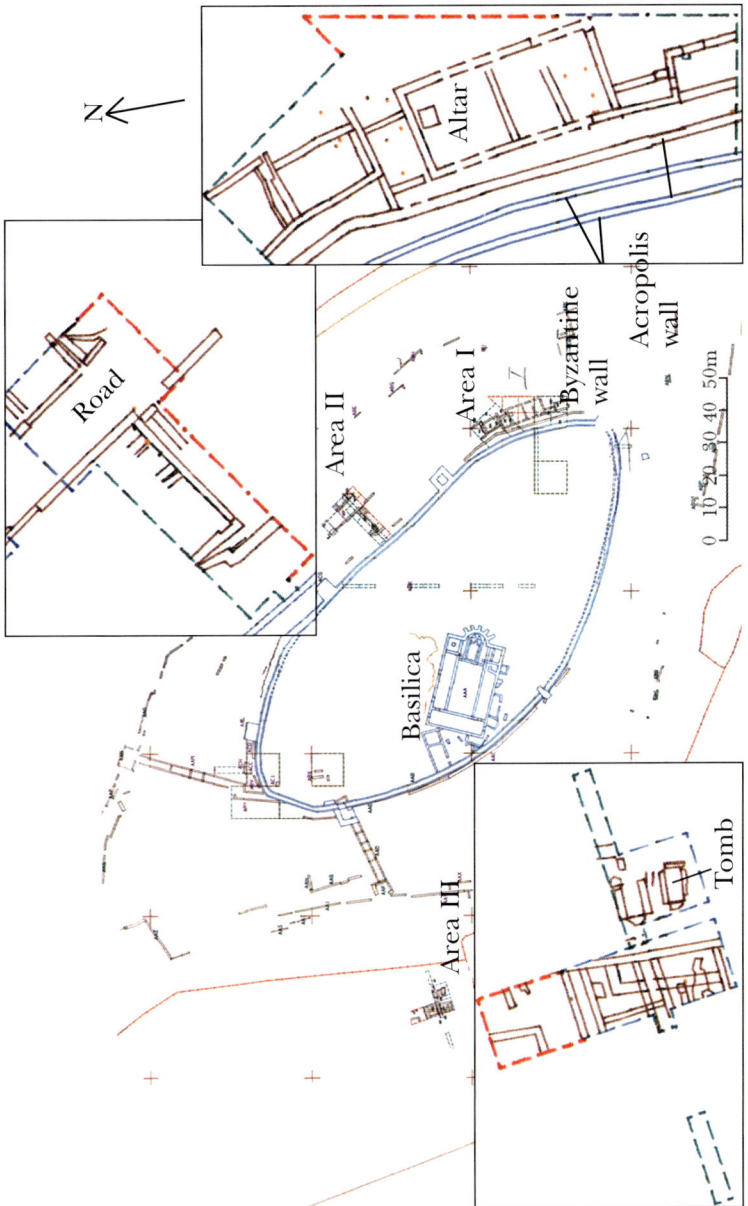

Plate 16. Haghia Triadha (Aetolian Chalkis): The trial trenches and units excavated in 1996-2000.

Plate 17. The substantial foundations for Archaic houses in the plain to the west of Aghia Triada, around trench 4. Seen from the south.

Plate 18. The Hellenistic house in excavation unit O26 is seen from the north towards the Gulf of Patras. To the right, the "Acropolis Wall" can be seen.

Plate 19. Hellenistic Cist grave in the plain, seen from the south.

Plate 20. A selection of finds from the Hellenistic Cist grave.

Plate 21. The excavated area seen from North.

Plate 22. Portico with cubicula, in the background a service corridor. The room in the middle was renovated several times. In the last phase with a medallion mosaic also featuring an artist inscription.

Plate 23. Peristyle. In the foreground a statue base situated in the axis of the peristyle. It is built into a small basin. In the middle, remains of the late Republican villa can be seen, and in the back is the eastern corner of the peristyle with a wide water channel in front.